Michael Rivkin, Ph.D.

BULK MATERIAL HANDLING

PRACTICAL GUIDANCE FOR MECHANICAL ENGINEERS

Library of Congress Control Number: 2018947038

ISBN
978-1-5437-4641-9 (sc)
978-1-5437-4642-6 (e)

Print information available on the last page.

To order additional copies of this book, contact
Toll Free 800 101 2657 (Singapore)
Toll Free 1 800 81 7340 (Malaysia)
www.partridgepublishing.com/singapore
orders.singapore@partridgepublishing.com

09/14/2018

PARTRIDGE

BULK MATERIAL HANDLING

PRACTICAL GUIDANCE FOR MECHANICAL ENGINEERS

Michael Rivkin, Ph.D.

"The Material Handling is the art and science associated with providing the right materials to the right place, in the right quantities, in right condition, in the right sequence, in the right orientation, at the right time and at the right cost using the right methods".

Material Handling Institute of America, Charlotte, NC, USA.

CONTENTS

INTRODUCTION

Bulk material handling is the sphere of mechanical engineering that is concentrated on design, operating, and maintenance of the equipment used for transporting bulk materials such as ore, rock, coal, grain, cereals, fertilizers, wood chips, so on.

The peculiarity of material handling is the fact that there are numerous technical solutions to any problem (parts of these are presented further on in this volume) and the component of the engineer's *personal* selection of the optimal solution is as critical as the *technical* component!

Today thousands of mechanical engineers are engaged in the design, upgrading and optimization of various material handling facilities.

This Guidance was written with the intention of helping the mechanical engineer who does not have a special background in material handling, to understand the main physical principles and the fields of application of various material handling systems, to familiarize the engineers with the *interdependence* between the equipment, used to handle bulk materials, and the parameters of the bulk material. This knowledge is required if one is to choose mechanically efficient and environmentally friendly equipment.

The analysis of the equipment, noting both the advantages and disadvantages of each type, is based on the personal experience of the author and his parsing of available technical information.

Presented in the Guidance, this information will help to the engineer to upgrade and optimize the existing bulk material handling facility or to select the proper equipment at the preliminary stages of a new project.

Having the essential technical information gathered from various sources, including the author's personal experience (decades in design, construction, upgrading, optimization, troubleshooting, and maintenance in Russia, Israel, Spain, and the USA), makes it easier to choose the optimal solution for one's specific technical problems.

GENERAL

In principle, bulk material handling includes the transportation, storing, distribution of, loading and unloading of bulk materials using the equipment that is the best technical solution for the given bulk material, for the required capacity, and for the local climatic conditions—environmentally friendly, and at the lowest possible cost.

Bulk material handling systems are the vital components of sea and river bulk ports, underground and open-pit mines, chemical plants, truck and wagon loading and unloading facilities, quarries, and so forth.

Ore, the raw bulk materials coming from mines and other sources, are very often land transported on overland belt conveyors.

Sea bulk carriers, barges, and trains are the most cost-saving means of transporting millions and millions of tons of various bulk materials over extended distances, so the focus of this Guidance is on the port material handling terminals where all these freights are received, stored, distributed and exported or imported.

Thousands of mechanical engineers operate, maintain, upgrade, and optimize the systems. In the USA in 2006, more than six hundred thousand technical specialists were employed in the various fields of bulk material handling.

A typical problem the mechanical engineer meets with is how to transfer bulk material from point A to point B. It can be done by use of belt conveyor (Conventional belt conveyor? Pipe conveyor? Flexowell conveyor? Cable Belt/Metso conveyor? RopeCon conveyor?), pneumatic conveyor (Dense phase? Dilute phase vacuum or positive pressure?), screw conveyor, drag chain conveyor, perhaps vibrating conveyor? There are so many technical solutions for any material handling problem that the most difficult and the most interesting task for the mechanical engineer is to choose the optimal technical solution.

The widely accepted definition of "the optimal solution to a technical problem" does not exist yet, but we can define it thus: "The optimal solution to a material handling problem is the simplest, most reliable, most easily implemented, most environmentally friendly and cost-effective solution. During technical discussions, no one from among your colleagues or your opponents (bulk material handling specialists) will be able to come up with a better solution than this".

Non-optimal solutions to material handling problems can cause technical and operational problems. Many new bulk material handling systems meet only 40%÷80% of capacity requirements in the first years of operation because of improper selection of storage and handling equipment and/or an inaccurate evaluation of the characteristics of the bulk material.

The choosing of the optimal material handling equipment (or handling system) is always a process based on the search and analysis of many options using knowledge, experience, an understanding of the physics of conveying systems, and creativity (yes, creativity!) on the part of the mechanical engineer.

Whether the chosen equipment will either succeed (i.e., be profitable, productive, and reliable, requiring minimal maintenance) or fail depending on the engineer's choice.

So, the devil is in the engineer's choice!

Bulk Material Handling Systems

Bulk material handling systems consist of the following:

- stationary machinery (belt conveyors, screw conveyors, elevators, pneumatic conveying systems, etc.)
- storage facilities (Dome-type storages, sheds, silos, covered piles, open piles, etc.)
- various mobile equipment (stackers, reclaimers, shiploaders and ship unloaders, grab cranes, etc.).

The properties of the bulk material play a dominant role in the choosing of the optimal handling equipment.

CHAPTER 1
Properties of Bulk Materials

As it was noted in the foregoing section, the properties of a bulk material have an appreciable effect on the choosing of the right handling equipment.

Particle size distribution, particle sizes d_{10}, d_{50}, d_{90} (Figure 1, Figure 2), flowability, hygroscopicity (absorption of moisture from the air), consolidation/agglomeration as a function of storage or transportation period, abrasiveness, corrosiveness, and so on, are the important characteristics that should be used as the data upon which to base the choice of the equipment. Requirements to prevent attrition and degradation of particles during the conveying, the loading and the unloading, prevention of dust emission, exert an effect on the choice of the proper material handling equipment.

The material handling equipment that successfully operates with one type of bulk material may be a complete failure when it is used to handle a different type or grade of material.

Climatic conditions are the factor that often plays the decisive role in the choosing of the optimal bulk handling equipment. For example, the coarse potash produced by Dead Sea Works Ltd (ICL) and delivered to the port of Eilat and the port of Ashdod for export acts like two different bulk materials. The potash in Eilat is usually free-flowing, non-corrosive, non-sticky bulk material, whereas the potash in Ashdod is non-free-flowing and corrosive bulk material.

The reason for the different behaviour of the potash is the difference in the moisture content of the material, which in turn depends on the relative humidity (RH%) of the environment. In Eilat the RH is 20%÷30%, and in Ashdod it is 85%÷95%. The potash, a hygroscopic bulk material, begins absorbing moisture from air at an environmental RH% greater than 65%. The moisture content of the potash in Ashdod can increase from an initial 0.1%÷0.3% to 0.5%÷2% (depending on the duration of storage and on the season), whereas in Eilat the potash maintains a low moisture content. Thus, high or low relative humidity can significantly change the characteristics of the hydrophilic bulk material.

For comparison, phosphate, which is a hydrophobic bulk material, does not change in characteristics at high relative humidity of the port of Ashdod.

Figure 1. Example of agglomerated bulk material.

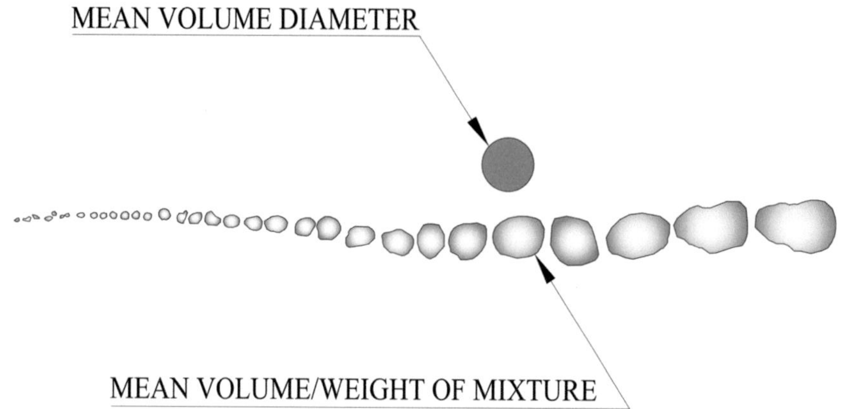

Figure 1.1 Graphic presentation of mean/average particle diameter.

The mean diameter of particle d_v (Souter's diameter) is the diameter of the sphere with the same volume/ weight as the real mean particle, whereas V_p is the volume of the mean particle:

$$d_v = \left(\frac{6V_p}{\pi}\right)^{1/3}$$

The average particle diameter of a mixture of particles in the sample can be calculated:

$$\overline{D}_3 = \frac{\sum_{i=1}^{N} N_i \overline{D}_{pi}^3}{\sum_{i=1}^{N} N_i \overline{D}_{pi}^2}$$

where N_i is a number of particles per each fraction.

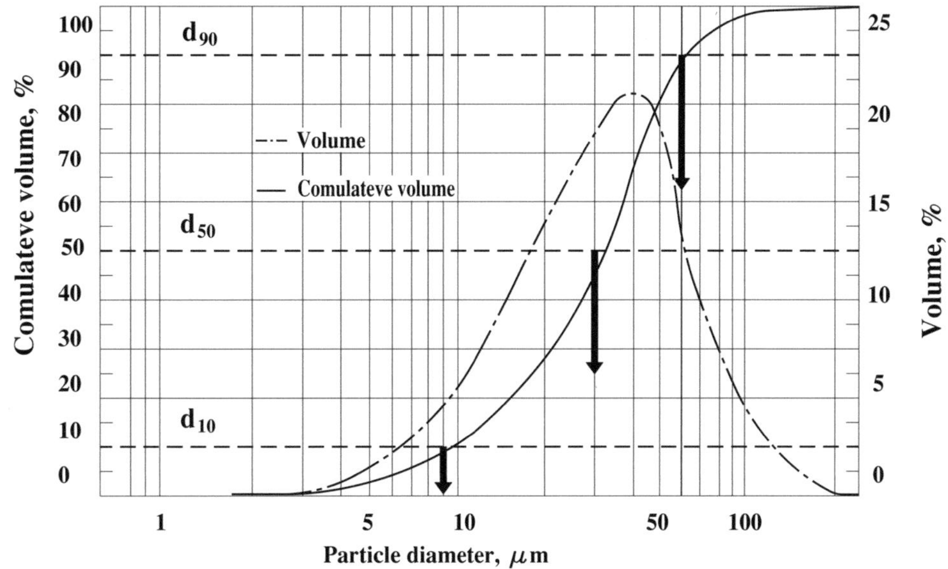

Figure 2. Graphical presentation of particle size distribution (volumetric and cumulative).

From the point of view of the practical material handling engineer, the mean diameter of a particle d_v does not add much valuable information. The useful data can be derived from d_{10}, d_{50}, d_{90} values and from the percentage of the fine fraction (–325 Mesh Tyler).

The flowing parameters of bulk material can be obtained from shear tests following Jenike's methods [1], showing how the *strength* of consolidated bulk material at the free surface depends on the *stress* (Figure 2.1).

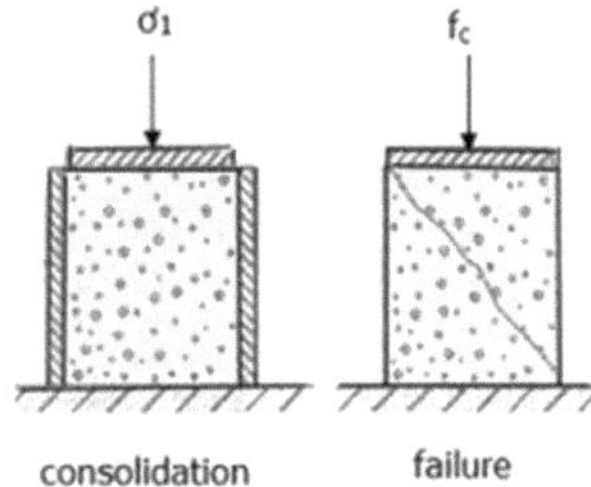

Figure 2.1 Experimental determination of strength of bulk material.

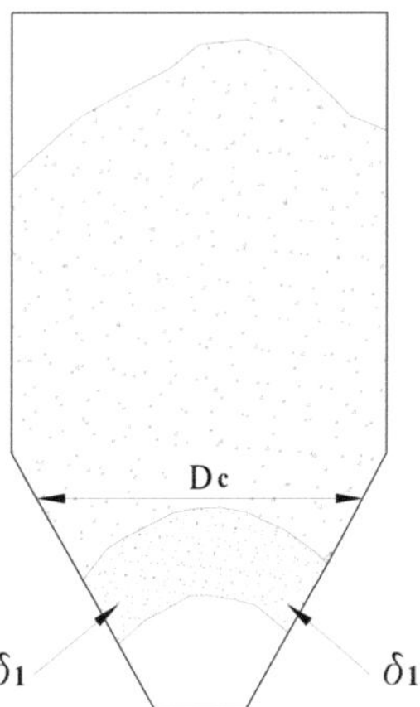

Fig. 2.2 The material stable arch or bridge at the outlet of the silo.

The definition of the angle of the lower cone of silo and the dimensions of the outlet are both based on the experimental data of the cohesive strength of the tested bulk material.

The critical diameter of the silo outlet D_c (Fig. 2.2) provides the free discharge of the bulk material (mass flow) when the bearing *stress*, acting on the arch of material σ'_1, is higher than the unconfined yield *strength* of material σ_c:

$$\sigma'_1 > \sigma_c$$

The conditions which the silo may be discharged by gravity is defined by the Flow Function (FF) [2]:

$$FF = \sigma_1 / f_c$$

- Very free flowing FF > 10
- Free flowing 10 > FF > 4
- Average flowing 4 > FF > 2
- Non-flowing, sticky FF < 2

The Code of Bulk Material [3] can be found in Table 1.

Table 1.

Code of Bulk Material

Miscellaneous Characteristics (Sometimes more than one of these characteristics may apply)	Very dusty Aerates and develops fluid characteristics Contains explosive dust Contaminable, affecting use or saleability Degradable, affecting use or saleability Gives off harmful fumes or dust Highly corrosive Mildly corrosive Hygroscopic Interlocks or mats Oils or chemical present—may affect rubber products Packs under pressure Very light and fluffy—may be wind-swept Elevated temperature	L M N P 0 R S T U V W X V Z

	Material Characteristics	Code
Size	Very fine—100 mesh and under Fine—1/8 inch and under Granular—Under 1/2 inch Lumpy—containing lumps over 1/2' inch Irregular—stringy, interlocking, mats together	A B C D E
Flowability Angle of Repose	Very free flowing—angle of repose less than 19° Free-flowing—angle of repose 20° to 29° Average flowing—angle of repose 30° to 39° Sluggish—angle of repose 40° and over	1 2 3 4
Abrasiveness	Nonabrasive Abrasive Very abrasive Very sharp—cuts or gouges belt covers	5 6 7 8

For example, very fine, free-flowing, non-abrasive, hygroscopic bulk material would be designated as Class A25U.

The definitions of bulk materials (coarse, fine, etc.) are linked with the particle size range:

- coarse bulk material, size range of 5 mm to 100 mm (ore, coal)
- granular bulk material, size range of 1.0 mm to 5.0 mm (granular potash)
- coarse powders, size range of 300 µm to 1.0 mm (coarse potash)
- fine powders, size range of 10 µm to 100 µm (fine phosphate, cement, apatite)
- superfine powders, size range of 1.0 µm to 10 µm (special powders)
- ultrafine powders, size range of < 1.0 µm (special powders).

How to eliminate or significantly reduce dust emission is one of the main problems of material handling. The initial load of dust can be 50 g/m^3 to 70 g/m^3 and more, as today the acceptable level of emission from dedusting systems is less than 20.0 mg/m^3 of TPM (Total Particulate Matter).

There are two interacting methods that may be used to reach the acceptable level of dust emission according to the environmental authorities:

1. A proper design that should be directed toward reducing the velocity at which the material falls and toward eliminating the impact of the bulk material transferred from one belt conveyor to another conveyor, being poured out into the hold of a ship or when loading railway wagons, trucks, etc.
2. Designing and installing corresponding dedusting systems.

The definition of dust is shown below:

Dust Classification

It is generally accepted to refer to particles $d < 47$ µm (–325 Mesh Tyler) as dust.

According to [4], the dust and the load of dust in the air can be presented as:

- Fine dust (50% < 5.0 µm) and light load of dust (< 70 g/m^3).
- Medium dust (50% from 5.0 µm to 15 µm) and moderate load of dust (70 g/m^3 to 175 g/m^3).
- Coarse dust (50% > 15 µm) and heavy load of dust (> 175 g/m^3).

Air pollutants $PM_{2.5}$ and PM_{10} (usually PM_{10} includes 50%÷70% of $PM_{2.5}$) consist of particles small enough to cause serious respiratory health problems (according to the World Health Organisation).

The flow rate of the dust emission to be removed by a dedusting system from various sources (transfer points, discharge pits, etc.) can be found in [4].

CHAPTER 2

Choosing of the Optimal Material Handling Equipment

Belt conveyors are the first and preferable option to be examined when seeking to find the right equipment for transporting various bulk materials.

Essential information about all aspects of conventional belt conveyors for bulk materials (calculations, design, etc.) can be found in [3].

In this Guidance, we will touch only the principles of operation and the practical aspects of different belt conveying systems, analysing the advantages and disadvantages of each system.

2.1 Belt Conveyors for Bulk Materials

2.1.1 General Description of Belt Conveyors

Belt conveyors [3, 5, 6, 7] are the most commonly used means of transporting bulks materials. This is because they are the simplest, the most reliable, require the least maintenance, and are the most investment-effective of the carriers of various bulk materials at any required capacity, over distances from a couple of metres (belt feeders) to tenths of kilometres (overland conveyors), and at the lowest power consumption in [kW/(tph × m)] (Table 2).

Thanks to the versatility, reliability, and low-maintenance requirements, belt conveyors are the key equipment for industry, mining, ports, and wagon- and truck-loading/unloading facilities, so on.

Table 2

MATERIAL HANDLING EQUIPMENT	POWER CONSUMPTION kW/ [tph x m]	METAL CONSUMPTION kg/ [tph x m]
BELT CONVEYOR	1	1
BUCKET ELEVATOR	2-3	6-8
EN-MASSE CHAIN CONVEYOR	3-4	5-7
SCREW CONVEYOR	4-5	3-5
PNEUMATIC CONVEYOR	8-15	1.5-2

A typical belt conveyor consists of a drive unit, an endless belt, carrying and return idlers, a supporting frame, tail and bend pulleys, a take-up system, belt cleaners, etc. (Figure 3).

Figure. 3 and Figure 4 represent the schemes of conventional belt conveyors.

Figure 5, Figure 6, and Figure 6.1 demonstrate operating belt conveyors.

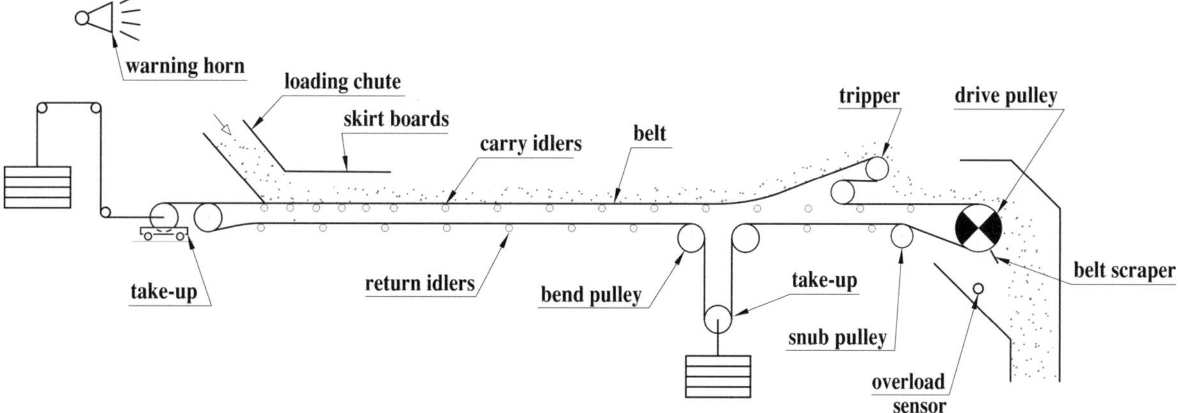

Figure 3. Principle scheme of conventional belt conveyor.

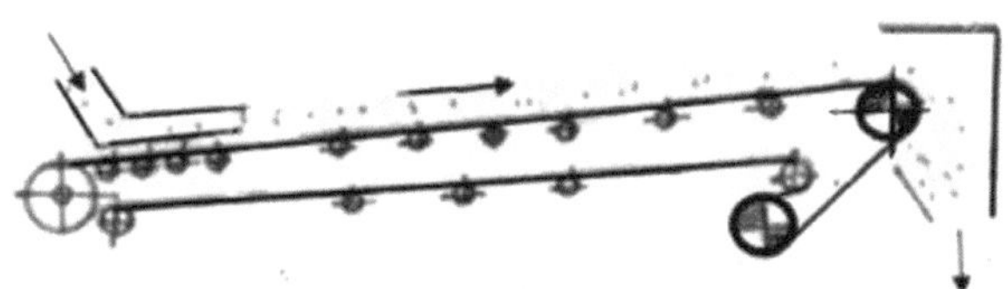

Figure 4. Scheme of dual-drive belt conveyor.

Figure 5. Belt conveyor in operation.

Figure 6. Steel cord belt (SCB) overland conventional conveyor (BEUMER Group).

Figure 6.1. The curved BEUMER trough belt conveyor in a tunnel.

There are many different types and configurations of conventional belt conveyors [3]. Only a small part of them shown in Figure 7 and Figure 8, where:

1. Belt feeder (usually up to 30 m long) with screw-type take-up on the head pulley and with a fixed drive tail pulley.
2. Long belt conveyor with double-drive unit located close to the tail section of the conveyor.
3. Overland conventional SCB conveyor with boosters.
4. Bidirectional belt conveyor.
5. Shuttle-type belt conveyor.
6. Long belt conveyor with double-drive unit located at the discharge section.
7. Modification of the double-drive unit located at the discharge section of the conveyor.
8. Long, double-drive belt conveyor with one drive unit located on the head section and with a second drive unit located on the tail section of the conveyor.
9. Belt feeder with drive head pulley and screw-type take-up installed on the tail pulley.
10. Belt feeder with the tail drive pulley used also as a screw take-up pulley.
11. Belt conveyor with drive head pulley and gravity take-up.
12. Belt conveyor/open pile loader, built and installed in DSW's Sdom plant in 1984, but could not be put into operation. It was re-designed later with modifications, reequipped (see no. 16) and has been successfully operating ever since.
13. Belt conveyor with drive unit in the middle of the conveyor and with tail screw take-up.
14. Bidirectional belt conveyor.
15. Belt conveyor with drive head pulley and with a carriage-type gravity take-up system.
16. Belt conveyor with drive pulley and carriage-type gravity take-up installed on the return belt in the middle of the conveyor.

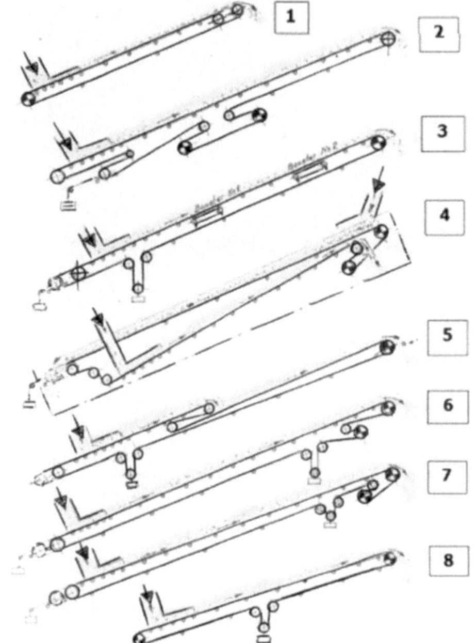

Figure 7. Different configurations of belt conveyors.

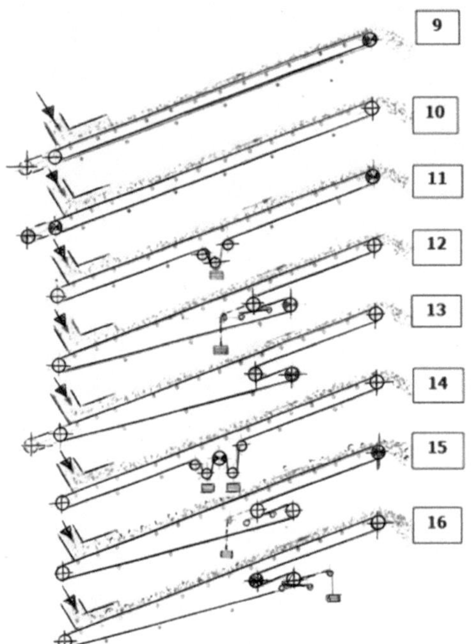

Figure 8. Different configurations of belt conveyors (continued).

Euler's law (see section 2.1.4) provides three possibilities to prevent belt/drive pulley slippage:

- increase the coefficient of friction between the belt and the pulley (μ).
- increase the wrap angle (α)
- increase take-up tension (T_2).

Euler's law also shows that the sufficient take-up tension (T_2) should be applied to the belt *after* the belt is wrapped around the drive pulley. This arrangement is needed to transfer (without slippage) the required hauling tension to the belt via friction between the drive pulley and the belt.

The result of the slippage is the rapid wear of the belt, accompanied by the dangerous heating of the belt, by overload of feed section of the conveyor and a complete stop of the conveyor operation.

As an example, years ago, one of DSW's belt conveyors was designed and manufactured with a take-up gravity system installed *before* the drive pulley (see no. 12, Figure 8). As a result, the non-stop slippage between the loaded belt and the drive pulley prevented normal operation of the conveyor. The conveyor was shut down, re-designed with a take-up installed *after* the drive pulley, reequipped (see no. 16), and has been successfully operating ever since.

All belt conveyors presented above (except for no. 12, which is shown before modification) are in operation now at different material handling facilities.

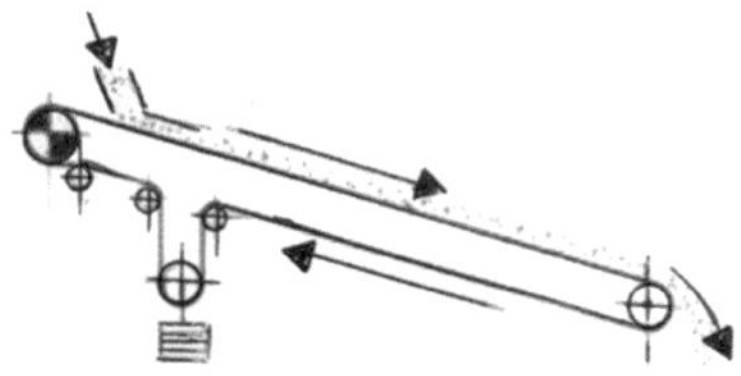

Figure 8.1. Downhill belt conveyor.

The scheme above (Figure 8.1, the case of lowering with a regenerative load) shows an uncommon case where the gravity take-up system can be located *before* the drive pulley ($T_2 = T_o + T_b + T_{yr}$ or $T_2 = C_w \times T_e$—use the larger value of T_2 [3]).

This downhill conveyor operates at such a configuration because the electric motor of the conveyor drive unit can operate at the braking (and not at the normal, driving) regime.

2.1.2 Technical Data

The general technical data required for belt conveyor calculations is presented below:

Main Parameters of the Bulk Materials

- Bulk density
- Particle size distribution (d_{10}, d_{50}, d_{90}, % of −325 Mesh Tyler)
- Flowability (free flowing, average flowability, non-flowing, sluggish)
- Repose angle, surcharge angle, internal friction angle
- Abrasiveness
- Corrosiveness
- Hygroscopicity
- Temperature of the loaded bulk material
- Additives to the bulk material (mineral oil, Armeen – anti-caking agent, etc.).

Local Conditions

- Relative humidity (summer, winter, maximum/minimum)
- Number of rainy days per year
- Ambient temperature (maximum/minimum, average).

General Data

- Conveyor length/route, lift
- Conveyor capacity
- Recommended conveyor speed
- Environmental requirements.

2.1.3 Main Safety Requirements

A brief list of main safety requirements for belt conveyors is shown below. Detailed information can be found in [8, 9, 10, 11]:

1. Maintenance must only be carried out by skilled and well-trained personnel.
2. Where no walkways have been provided, an approved lifting platform must be used.
3. Emergency pull-wire rope, with stop switches every 30 m÷50 m, must extend the entire length of a conveyor on both its sides.
4. Return idlers should be guarded to prevent their fall if maintenance personnel cross under, work under or clean under operating belt conveyor (Figure 9).
5. Nip points and open rotating parts of the drive units, take-up system, and tail section should be covered with safeguards (Figure 10).
6. The process for removing and reinstalling guards must be safe, quick and easy.
7. The weight of each removal guard must be less than 20 kg.
8. Access from above the guards should be prevented.
9. The barrier guards should surround danger zones (e.g. a 2-metre-high guard around gravity take-up weight, with 300 mm clearance from ground level).
10. No loose clothing must be worn to avoid entanglement in the machinery.
11. Safeguards should be provided even at the design stage of the project.

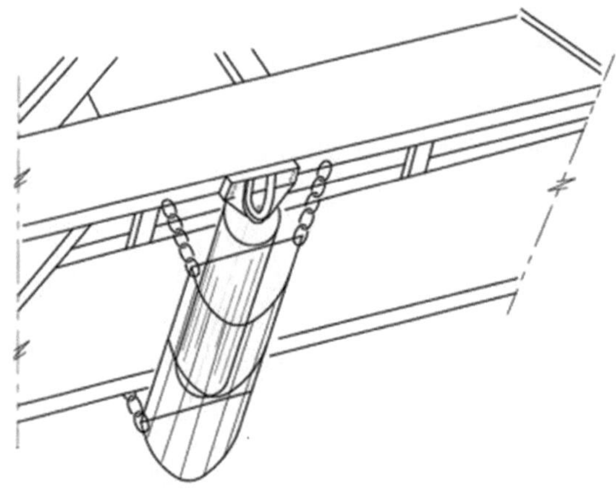

Figure 9. Special guard for return idlers.

To prevent possible accidents during regular inspection and maintenance or by accidentally putting one's hand in during conveyor operation, the nip points of conveyors must be guarded (Figure 10, Figure 10.1). The guards are to be fixed to the pulley support frames [8, 9,10, 11]. Guard dimensions: from 60 mm × 60 mm (for span < 0.6 m) to 80 mm × 80 mm (for span > 1.0 m). Guards are not required for bend pulleys.

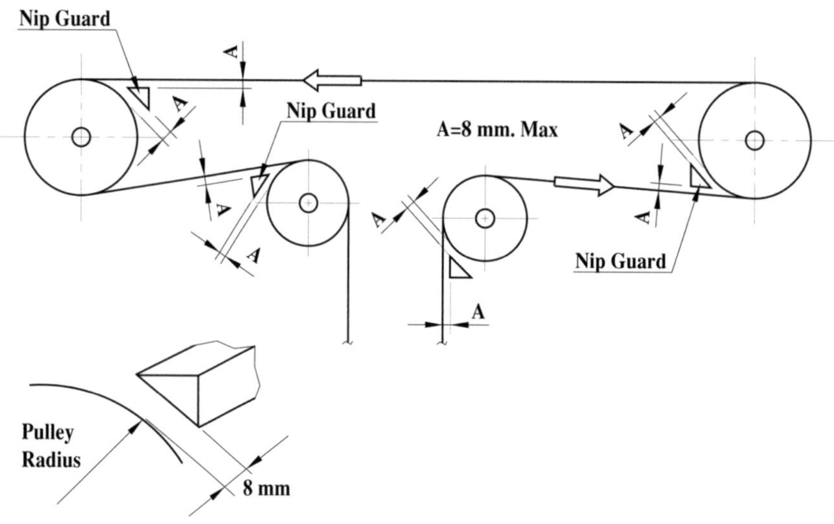

Figure 10. Principles of guarding of nip points.

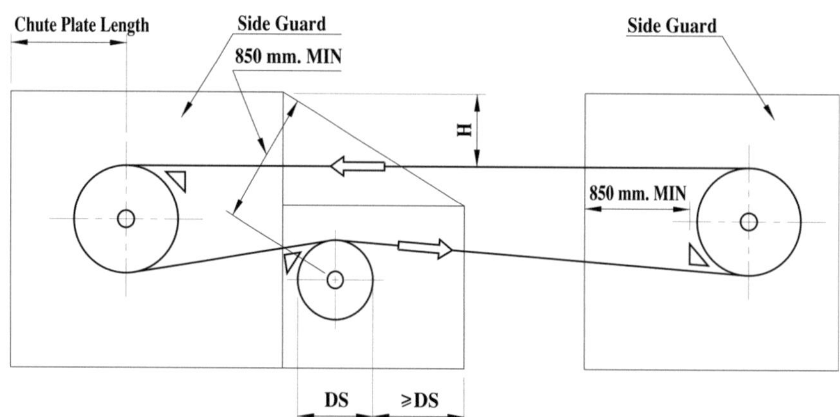

Figure 10.1. Guards for nip points and for open rotating parts of belt conveyor.

2.1.4 Calculations of Belt Conveyors

As noted above, the normal operation of a belt conveyor is based on the transferring (without slippage) of the required hauling force/tension to the conveyor belt via friction between the drive pulley and the belt wrapped around the pulley.

In principle, all calculations of conventional and overland belt conveyors [3, 6] are based on the solution of two equations: Euler's law (1) and the resistance summarizing equation (2):

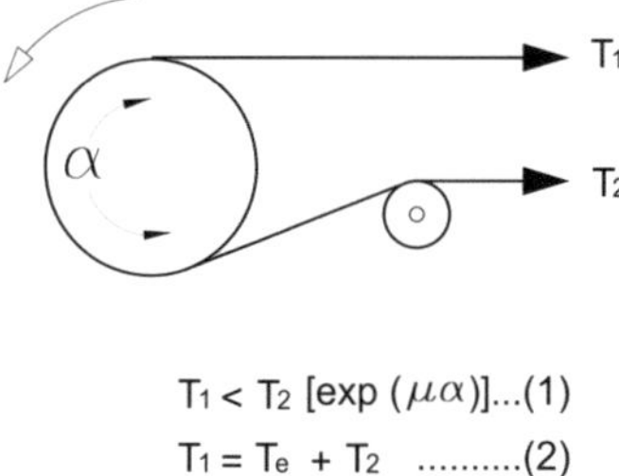

$$T_1 < T_2\,[\exp(\mu\alpha)] \ldots (1)$$

$$T_1 = T_e + T_2 \ldots\ldots\ldots (2)$$

where T_1 is the tight-side or hauling belt tension at the drive pulley; T_2 is the slack-side belt tension at the drive pulley, required to create the friction drag between the drive pulley and the conveyor belt, enough for the normal hauling of a loaded conveyor without slippage; $T_e = \Sigma\, T_{ei}$ is the effective tension at the drive. Summarized resistance tension is produced by:

T_{e1} – tension resulting from acceleration of material,

T_{e2} – tension resulting from lifting/lowering of material,

T_{e3} – tension resulting from pulleys resistance,

T_{e4} – tension resulting from resistance of carrying and return idlers,

T_{e5} – tension resulting from resistance of belt to flexure,

T_{e6} – tension resulting from resistance of material to flexure,

T_{e7} – tension resulting from friction of the skirt board,

T_{e8} – tension resulting from resistance of the belt cleaner,

T_{e9} – tension resulting from plough resistance.

α – wrap angle, rad

μ – coefficient of friction between the belt and the drive pulley

The conveyor calculations can be carried out using many of the available computer programs.

The example below demonstrates one of the simplified methods of belt conveyor calculations, developed by the author and based on the Conveyor Equipment Manufacturers Association (CEMA) [3].

2.1.5 Example of Belt Conveyor Calculations

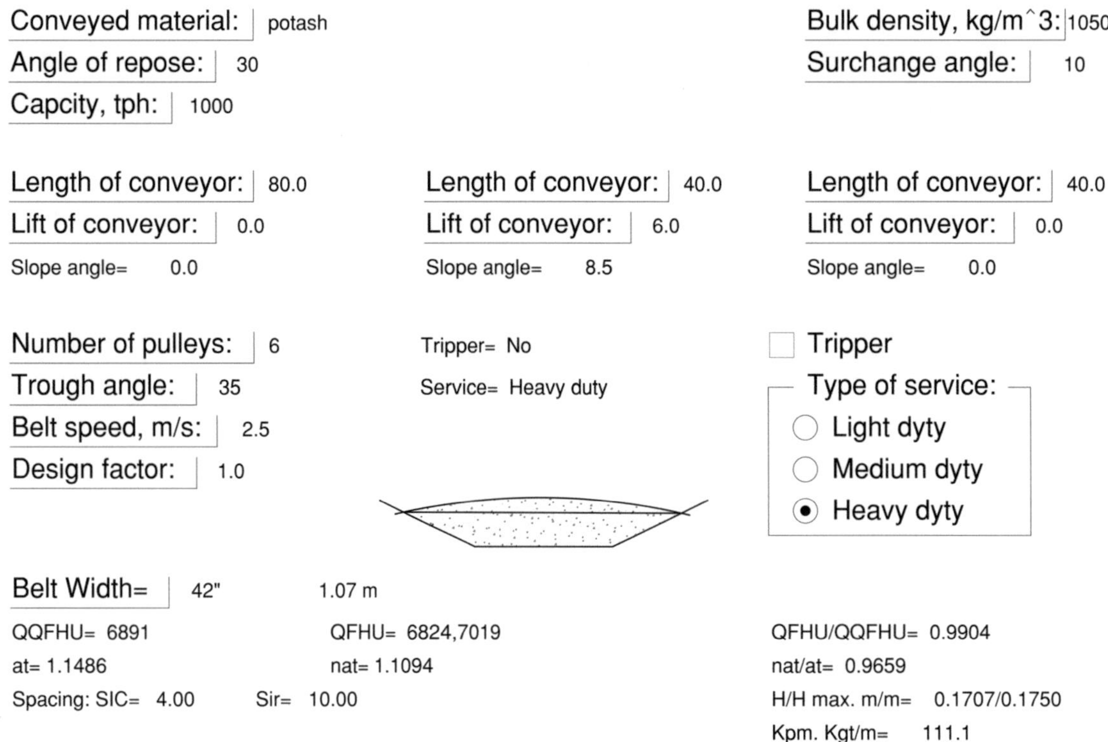

The program allows online change one or several technical parameters (bulk density of the material, surcharge angle, conveyor speed, trough angle, belt width, capacity, etc.) and re-calculates the conveyor in accordance with the new data.

The cross section of the conveyor is shown in scale. This helps us to visually estimate the conveyor loading.

The program also recommends installing (or not installing) a variable frequency drive (VFD) or a hydraulic coupling to perform a "soft start" during acceleration of the fully loaded belt conveyor.

Take-up (T_o, T_{2M}) Calculations

Length of skrtboards, m: 5

- (•) 2% Sag
- ( ) 3% Sag

Wrap Angle: 190

Calculation:

Friction coeffcient= 0.30
Wa= 150
TO= 2137
T2M= 2239

	Drive pulley	Conditions
( )	Lagged	Dry
(•)	Lagged	Semy-wet
( )	Lagged	Wet
( )	Bare	Dry
( )	Bare	Semy-wet
( )	Bare	Wet
( )	Other	

Distance between head pulley and take-up pulley: 10

Number of pulleys: 4

Results:

Material:	potash
Bulk density, kg/m^3:	1050
Angle of repose:	30
Surchange angle:	10
Length L1, m:	80.0
Height L1, m:	0.0
Slope angle L1=:	0.0
Length L2, m:	40.0
Height L2, m:	6.0
Slope angle L2=	8.5
Length L3, m:	40.0
Height L3, m:	0.0
Slope angle L3=	0.0
Capacity, tph:	1000
Belt speed, m/s:	2.5
Belt width:	42"
Construction of belt EP=	315/4
Drive pulley diameter, mm=	508

Trough angle:	35
Wrap angle:	190
Friction coef.:	0.30
Motor acceleration time ,s:	4.5
Use centrifugal coupling or soft start.	
Required HP:	67.5
Motor HP:	75
Take-up ,kg:	2707
Spacing ,m:	Sic=1.20 Sir=3.00
Srart torque ratio R=	1.3544
Safety factor=	1.1113

Results:

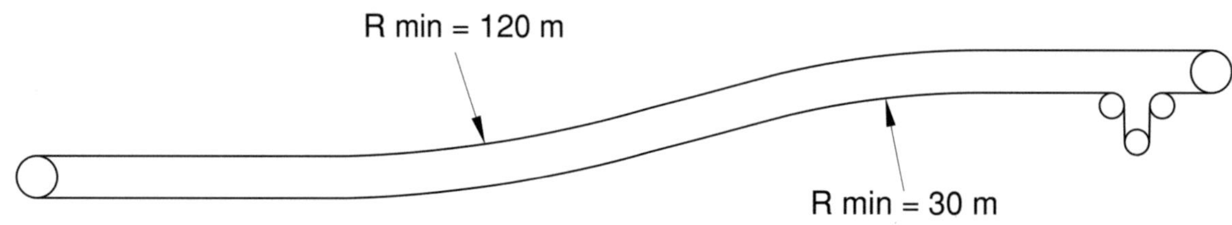

Graph: ordinates of concave vertival curve.

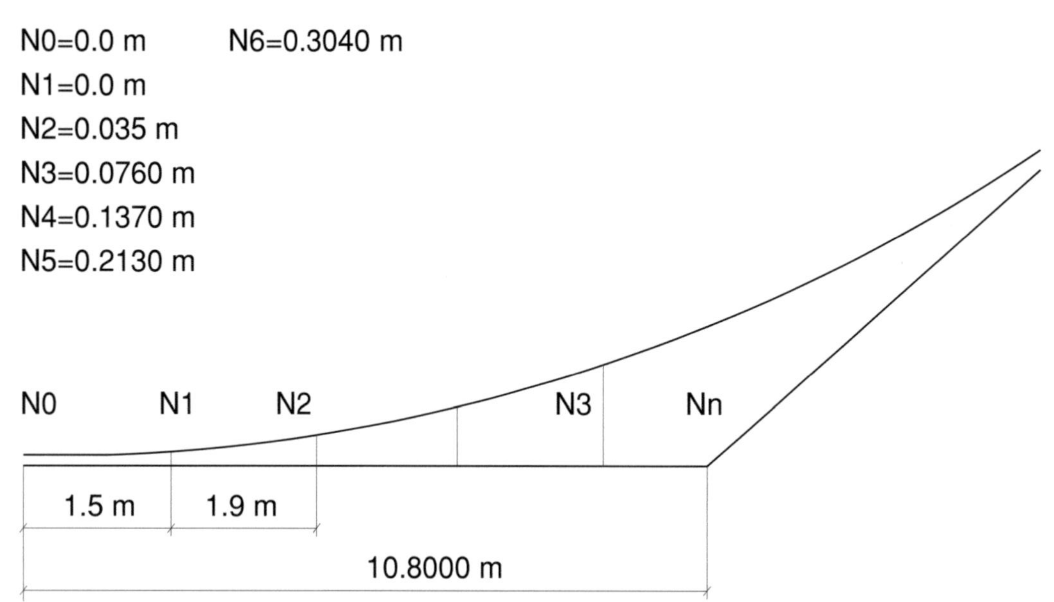

Belt tension diagram:

T1=3105 kgf

T2=1148 kgf

T0=970 kgf

TE=1957 kgf

2*TTT=2707 kgf

Trajectory of the discharged bulk material:

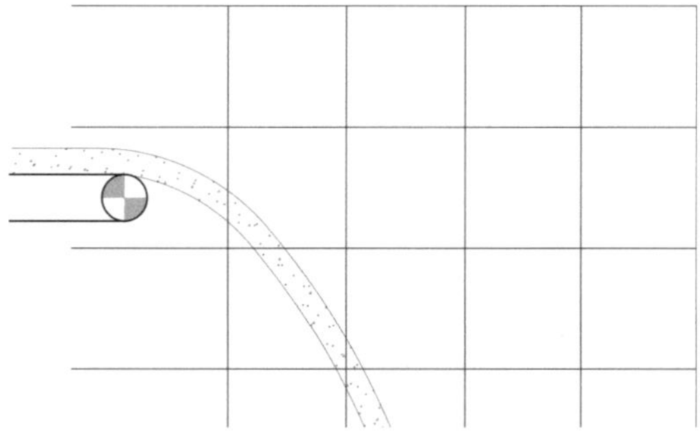

Conveyor feeding section:

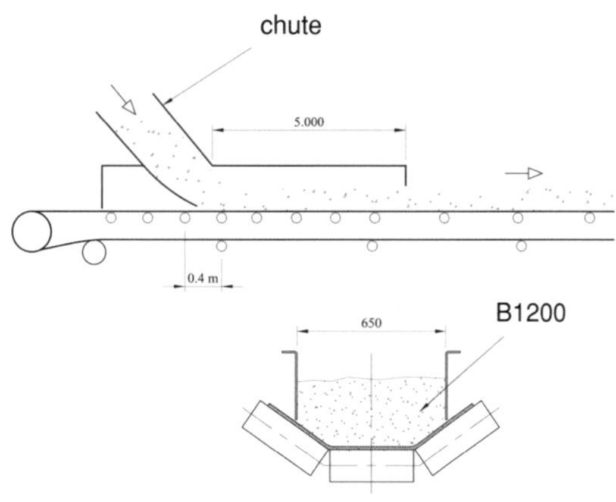

Calculations of the chute cross section:

The minimum cross section of a chute can be calculated (A. Fritella and A. Smit):

$$A = \frac{2.5 \cdot C_{dc}}{3600 \cdot S \cdot D} \ [m^2]$$

where S is the material flow speed; m/s, D is the bulk density; t/m³, and C_{dc} is the designed capacity, tph.

Conveyor pulleys:

Drive pulley diameter, mm: 508

Drive pulley shaft diameter= 4 1/4"
Shaft defection tangent tan a= 16.8012E-04
a= 16.8012E-04

Drive pulley shaft material:
- (•) AISI C1018
- ( ) AISI C1042
- ( ) AISI 4030

☐ Key Ringfeder

Take up pulley:

Take up pulley shaft diameter= 3"
Take up pulley diameter, mm= 400

Take up pulley shaft material:
- (•) AISI C1018
- ( ) AISI C1042
- ( ) AISI 4030

☐ Key Ringfeder

Bend pulley:

Bend pulley shaft diameter= 3"
Bend pulley diameter, mm= 300

Bend pulley shaft material:
- (•) AISI C1018
- ( ) AISI C1042
- ( ) AISI 4030

☐ Key Ringfeder

\- END -

2.1.6 Take-up Systems

Correct selection of the take-up tension T_2 is one of significant factors to maintain the belt conveyor normal operation by preventing the slippage (condition no. 1).

The tension T_2 should also ensure the chosen sag (usually, between 2% and 3%) of the loaded belt between two neighbouring idler sets (condition no. 2).

The larger of the two values of tension T_2 will be used for further calculations.

In most cases, the gravity devices are used to take up the conveyor belt. This simple device automatically provides the required take-up tension independently on the stretch of the belt. The screw take-up, recommended for short (up to 30 m long) conveyors/feeders, is usually mounted on the tail pulley and the stretch of the belt is regularly tightened by manually twisting the screw (Figure 12). In addition, this pull-up device can serve to adjust the conveyor belt.

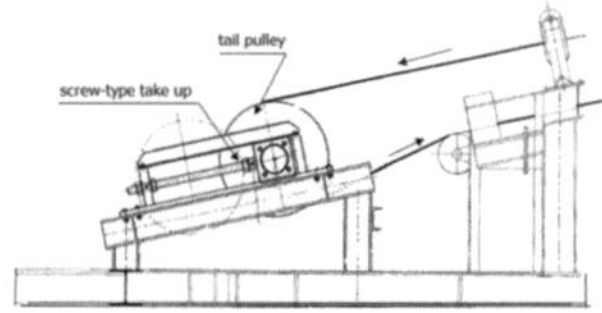

Figure 12. Typical screw-type take-up device installed on tail pulley.

Typical gravity take-up systems recommended for long belt conveyors are shown in Figure 13, Figure 14.

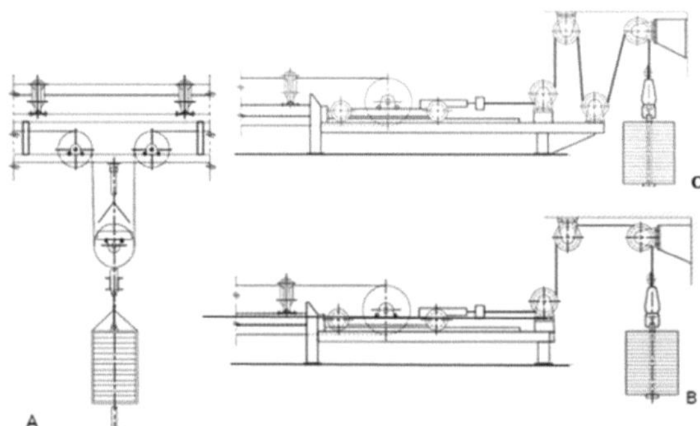

Figure 13. A-type and B-type gravity take-ups (C-type is the modified B- type).

The A-type is the common take-up device, installed, usually, close to the drive unit. Such an arrangement reduces the take-up weight and lessens the belt tension (Figure 13, Figure 14).

The modified A-type and B-type gravity take-ups (Figure 13 C and Figure 14.1) are used in cases when the lack of height *H* doesn't enable the required movement of the counterweight to take in the stretch of the belt.

The B-type tension unit is installed on tail pulleys of short belt conveyors/feeders. We would recommend that one install the screw units (Figure 12) on the tail pulleys of long belt conveyors - not as the take-up device, but for easy and safe belt adjustment (the adjustment can be carried out safely even when the conveyor is running).

Figure 14. Typical A-type gravity take-up system.

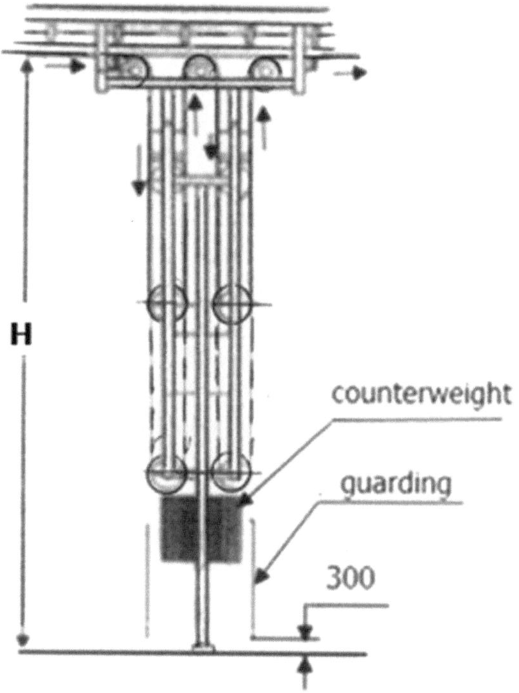

Figure 14.1. Modified A-type gravity take-up installed where the height (H) is limited.

The fully automatic take-up devices with take-up pulleys installed on the carriages and pulled by wire rope are often used in underground mines. One end of the rope is attached to the fixed structure via load cell, and the second end is wrapped around the drum of electric winch (Figure 14.2).

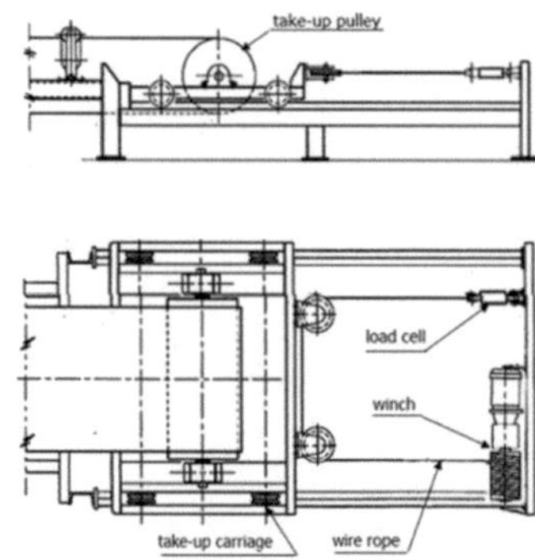

Figure 14.2. Take-up system used for underground belt conveyors.

Any decrease in rope (and, hence, belt) tension causes the signal from the load cell, actuating the winch, which starts to pull up the rope (and take-up carriage), thereby increasing the belt tension and getting it back to the required value.

Recurrent increase in rope/belt can be also carried out with manual winch (Figure 14.3), as maintenance people should periodically pull up the rope. Maximum capacity of standard manual winch is 30,000 N at hand force of 180 N.

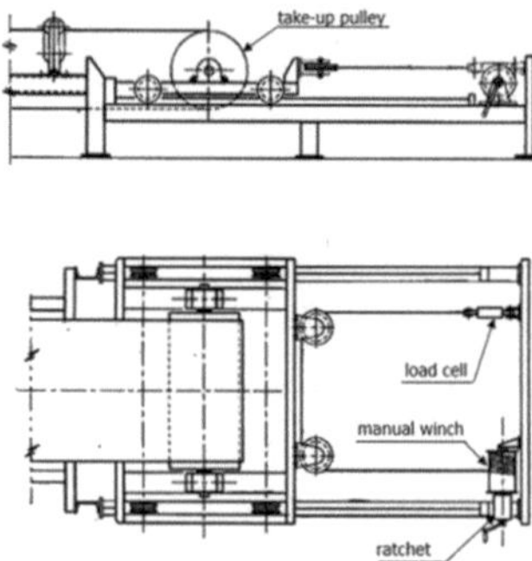

Figure 14.3. Horizontal take-up system with manual winch.

Such systems are recommended only for cases where the height restriction doesn't allow the use of a simple and reliable gravity take-up system shown in Figure 14 and Figure 14.1.

2.1.7 Belt Conveyor Galleries

To eliminate spillage and fugitive dust emission during conveying, Environmental Protection Agencies today require that outdoor belt conveyors for bulk materials be, at minimum, covered (Figure 15, Figure 15.1, Figure 16) and, as rule, enclosed in galleries (Figure 17, Figure 18, Figure 18.1, Figure 19).

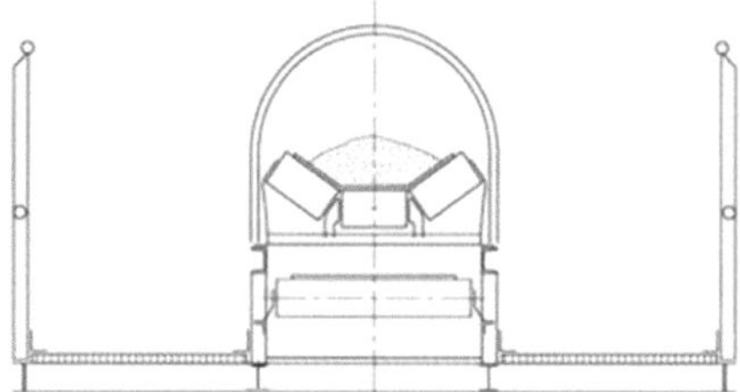

Figure 15. Covered belt conveyor with two-side walkways and sealed floor.

Figure 15.1. Example of covered belt conveyor.

Covered belt conveyors usually used to transport non-dusty bulk materials, or for transporting inside factories.

Enclosed galleries with a one-side walkway, shown in Figure 17, are slightly cheaper than galleries with two-side walkways, but they are non-optimal solution from the point of view of performing the maintenance because of difficulties of replacing the pulleys, carry and return idlers, of performing a splice, of replacing the belt, and so forth.

Figure 16. Covered belt conveyors.

Figure 17. Typical conventional belt conveyor gallery with one-side walkway (Neuero Group).

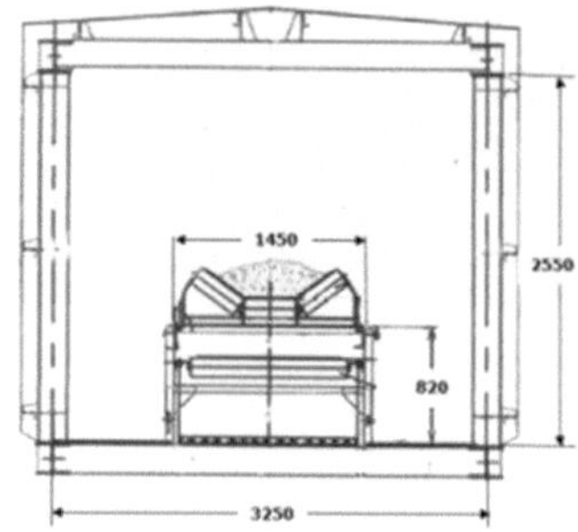

Figure 18. Conventional, truss-type gallery with two-side walkways for B1200 belt conveyor.

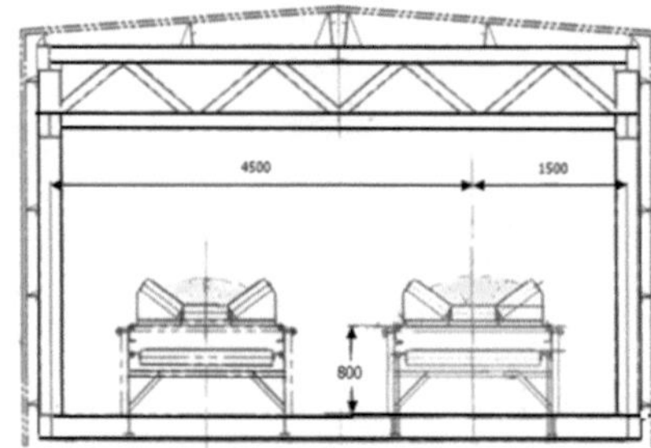

Figure 18.1. Typical enclosed conventional gallery for two conveyors (B1000 and B1200).

Figure 19. Example of fully enclosed conventional galleries for one conveyor and for two conveyors.

The *advantages* of fully enclosed galleries include the isolation of the conveyed bulk material from rains, winds, etc., and the prevention of dust emission and spillage during conveying. As noted above, to simplify inspection, maintenance and replacement of the equipment, it is recommended that the galleries will be designed with two-side walkways (see Figure 18, Figure 18.1).

Most of today's belt conveyor galleries are enclosed truss-type with sealed bottoms (Figure 19).

The alternative to the truss-type gallery is the steel pipe gallery, which is a tubular structure of 2.5 m to 3.5 m in diameter, installed above the ground with about 30-metre clear spans between supports.

For example, the two steel tubular conveyor galleries, 3.4 m in diameter, 120 m and 150 m long (for belt conveyors with a capacity of 750 tph), were installed during upgrading of the port of Eilat (Israel) in 2012–2013 (Figure 20).

These steel tabular galleries with two-side walkways, allowing installation of B1000 belt conveyors and also piping for compressed air and water supply, piping for internal vacuum cleaning, power supply, control, and lighting cables and lamps, were chosen as the optimal solution for the Eilat project and were approved by the Environmental Protection Authority. It might be interesting to note that conventional truss-type galleries were not approved for this project by the Authority due to the environmental restrictions.

Figure 20, Figure 21, Figure 22, and Figure 22.1 show the 3.4 m in diameter steel tubular conveyor gallery.

Figure 20. Two tubular steel belt conveyor galleries (3.4 m in diameter) linked in series.

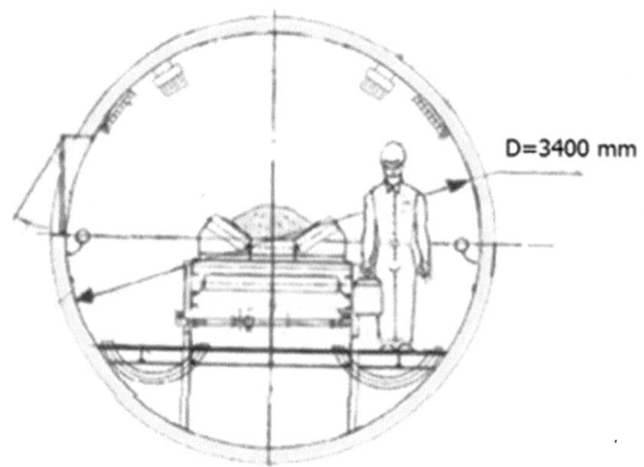

Figure 21. Cross section of the gallery shown in Figure 20.

Figure 22. This picture was taken inside of the steel tubular 3.4 m in diameter gallery shown in Figure 20.

Figure 22.1 The 3.4 m in diameter tubular gallery with the supporting frame and with emergency escape exit. The exits are located every 30 m for safety reasons.

The total weight of a 3.4 m in diameter tubular gallery carrying B1000 or B1200 belt conveyors is 750 kg/m to 800 kg/m, which is about the same weight (and, respectively, the same cost) as a completely enclosed conventional truss-type gallery carrying the same belt conveyor.

Advantages of a steel tubular gallery:

1. No dust emission.
2. Low-maintenance requirements.
3. The cost of the steel tubular gallery is very close to the cost of the conventional, truss-type, fully enclosed gallery.

Disadvantage of a steel tubular gallery:

The corrosive environment can limit the use of carbon steel made tubular galleries unless the external surface of the gallery will be painted with a corrosion-resistant paint.

Figure 23. Tubular belt conveyor gallery with one-side walkway (Bedeschi Mid-West Conveyor, USA).

Figure 23.1 Tubular belt conveyor galleries and pipe conveyors for limestone (Bedeschi Mid-West Conveyor, USA).

2.1.8 Drive Pulley Arrangement

Euler's law (section 2.1.4) revealed that effective transfer of friction tension/force from the drive pulley to the conveyor belt can be achieved by increasing the belt wrap angle (α) which accompanies the corresponding reduction in slack-side belt tension (T_2). To increase wrap angle, the drive pulley(s) may be arranged in various positions (Figure 24, Figure 24.1, Figure 24.2).

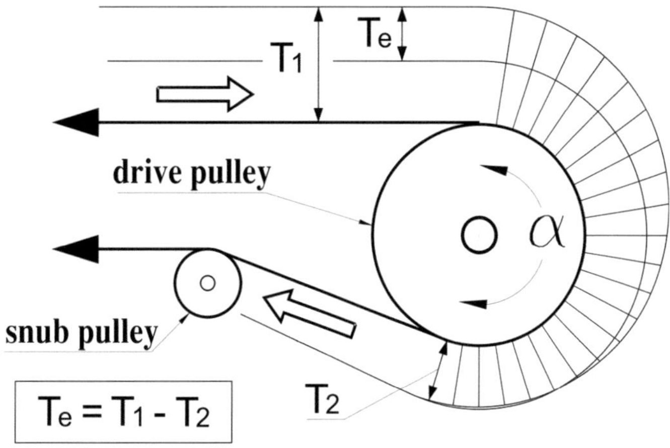

Figure 24. Distribution of belt tension around drive pulley.

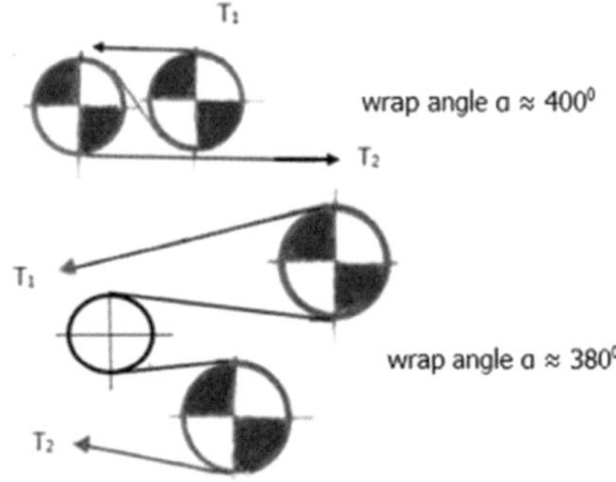

Figure 24.1 Various arrangements of double-drive units.

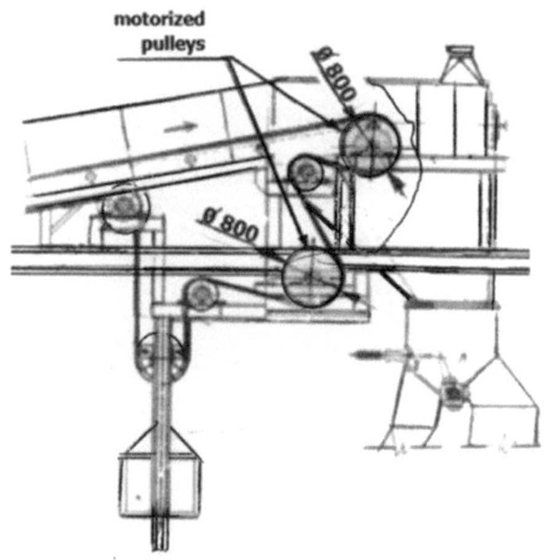

Figure 24.2 Design of double-drive unit with two 55 kW motorized pulleys.

In the case of using a double-drive unit, the calculations show that the wrap angles are summarized:

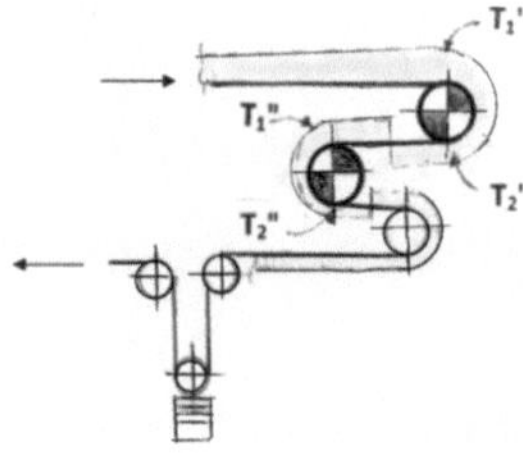

Figure 25. Scheme of calculations.

$$T_1'' < T_2'' \exp(\mu_1 \alpha_1),$$

$$T_1'' = T_2',$$

$$T_1' < T_2' \exp(\mu_2 \alpha_2),$$

therefore,

$$T_1' < T_2'' \exp(\mu_1 \alpha_1) \times \exp(\mu_2 \alpha_2),$$

or

$$T_1' < T_2'' \exp(\mu_1 \alpha_1 + \mu_2 \alpha_2)$$

or for $\mu_1 = \mu_2 = \mu$

$$T'_1 \leq T''_2 \exp[\mu(\alpha_1 + \alpha_2)]$$

Hence, the installation of the second drive pulley significantly increases the wrap angle, increases drive pulley/belt friction, and reduces the required take-up tension of the belt. The disadvantage of such a configuration is that the second drive pulley contacts the "dirty" side of the conveyor belt. Figure 24.2 shows how the use of an additional bend pulley helps to overcome this problem.

Most of the belt conveyors operating in underground mines, are designed with a double-drive unit arrangement (to increase reliability) and with total safety factor (SF) of 2÷2.5 (Figure 26).

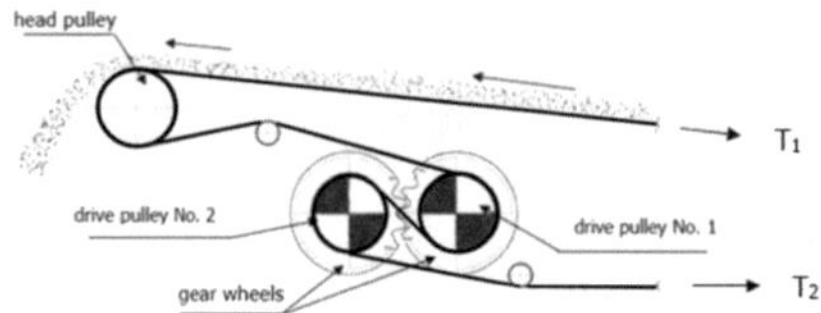

Figure 26. Typical drive system used in a mine underground belt conveyors.

The drive unit, shown in Figure 26, significantly increases the wrap angle and, accordingly, reduces the required belt tension T_2.

Advantages of this arrangement are:

1. Increase of wrap angle and, as a result, reduction of the required take-up tension.
2. The location of drive units is on the ground level facilitates the maintenance.
3. Reliability: the system can operate using one drive unit only.

Disadvantages of this arrangement are:

1. Complex and expensive three-pulley (one head pulley and two drives) system, with one drive pulley in contact with the "dirty" side of the belt.
2. Considerable tension ($2 \times T_1$) applied to the head pulley and to the discharge tower.
3. Spillage from overhead return belt on the drive unit.
4. The overdesign of the drive units leads to unreasonably high Opex and Capex.

Recommendation:

The transformation of the system into a one head pulley with two-side drive units (for reliability) can make the system more compact, simple, less expensive, and easy to maintain. The use of motorized pulleys is highly recommended.

2.1.9 Drive Unit Design: Conventional Drive Unit vs Motorized Pulley

The drive unit of a belt conveyor can be designed as a conventional drive unit (consisting of a head pulley with its outer bearings, reducer(s), motor(s), two couplings, special supporting frames, and fixed guards surrounding the rotating parts) or as a motorized pulley (holding the electric motor and gearbox inside a hermetically sealed, lubricant-filled shell, which protects against dust, water, and so on) (Figure 27).

Modern belt conveyors are mostly equipped with motorized pulleys, that have proven their high reliability and operate with little or no maintenance.

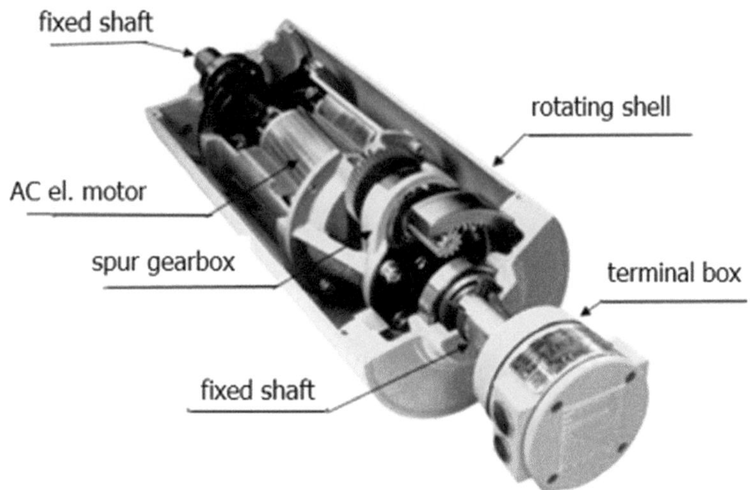

Figure 27. Typical motorized pulley (Rulmeca, Germany).

The comparison between footprints of a conventional drive unit and a motorized pulley is presented in Figure 28.

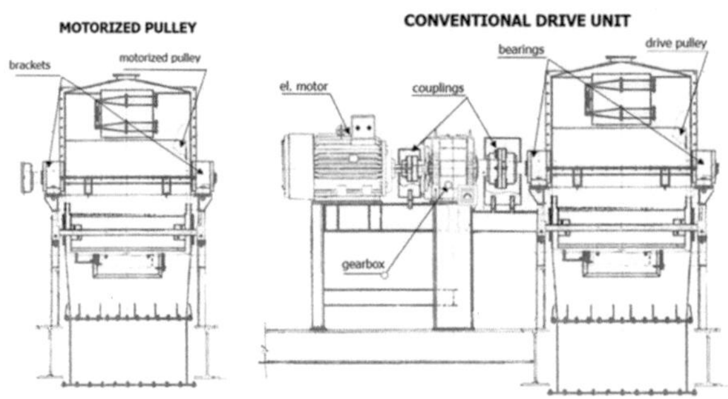

Figure 28. Comparison of footprints: motorized pulley vs conventional drive unit.

The additional advantage of motorized pulleys is the attractive cost. The cost of a motorized pulley (up to 75 kW) is lower by 10% to 20% than the total cost of a conventional drive unit.

Another advantage is the oil changes requirement after a very long period of operation. The mineral oil changes interval for motorized pulleys is about 20,000 hours of operation (for synthetic oil—every 50,000 hours !).

For comparison, the re-greasing interval for standard bearings of a conventional drive unit is about 1,000 hours at +70°C, but this is reduced by half every +10°C in temperature above +70°C.

The belt conveyor drive unit, included motorized pulley, belt cleaner, and heavy-duty flat idler (installed instead of a snub pulley) is shown in Figure 29.

Figure 29. The drive unit with motorized pulley.

2.1.10 Optimization of Drive Unit

A conventional drive unit consists of a head pulley, gear coupled with electric motor, snub pulley, belt cleaner, overload sensor and so on. The head pulley is installed within the upper hood that is bolted to the lower hopper.

The use of snub pulleys is the acceptable practice to increase the belt/drive pulley wrap angle. The snub pulleys are usually installed outside of the head hood (Figure 30).

During transport of dusty bulk materials, continuous spillage occurs when the return belt wraps the snub pulley that installed outside of the hopper. The spillage is accumulated under the snub pulley even if the belt is cleaned by a belt cleaner/scraper installed within the hood. Regular manual cleaning of the spillage is a time- and labour-consuming operations.

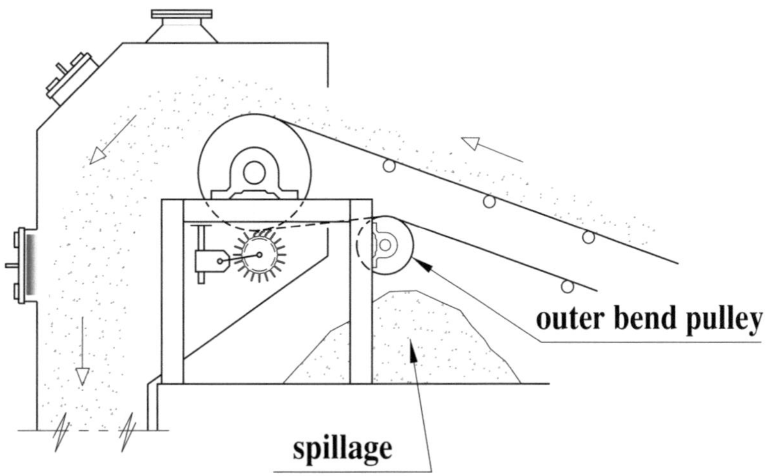

Figure 30. Traditional design: drive unit with outer snub pulley.

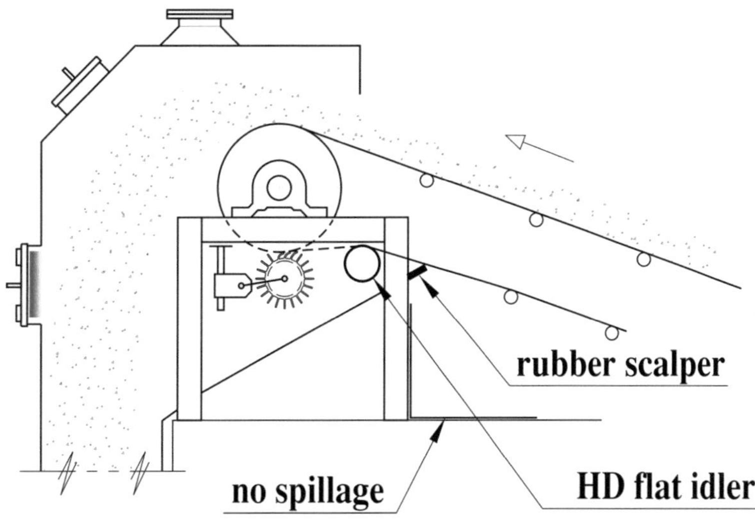

Figure 30.1 Optimized drive unit with motorized pulley and heavy-duty flat idler mounted inside the hopper (instead of the outer snub pulley).

Someone can suggest an option to install a snub pulley inside the existing lower hopper. It can be done, but ... the installing a snub pulley inside an existing hopper requires performing of significant (and often practically impossible) structural changes of the hopper. The width and the height of the lower hopper should be increased significantly and this will indisputably cause difficulties in transferring the material onto existing takeaway belt conveyor. In addition, the replacement of a broken snub pulley, located inside of the hopper, arises additional very serious difficulties for maintenance people.

To solve this problem on the existing belt conveyors at the port of Eilat, the snub pulleys on the belt conveyors were replaced by heavy-duty (HD) flat idlers, which were installed inside existing hoppers (Figure 30.1, Figure 30.2). The HD idlers were chosen because they can withstand the resulting force from the take-up tensioned belt wrapping the idler.

After it was established in practice that the spillage reduced drastically, the solution was implemented on a number of existing and new belt conveyors. The innovation, which has beten successfully operating since 2010, was recommended and accepted by the designers of the belt conveyors of the new port bulk handling terminals.

The optimized design of a conveyor head chute with a HD flat idler instead of the standard snub pulley is presented in Figure 30.1.

On the shiploader boom conveyor, where replacement of a pulley is a difficult and time-consuming operation, we also replaced the snub pulley with a HD flat idler where it successfully operated for many years.

Figure 30.2 Picture shows HD flat idler installed instead of outer snub pulley (port of Eilat).

The example of specifications of HD flat idlers that replaced snub pulleys, where they have been successfully operating at our facilities since 2010, are: PSV – 6 (ø159 × 4.5/ø35) and PS V –7 (ø159 × 4.5/ø40) at a permitted load capacity of 990 daN (Rulmeca, Italy).

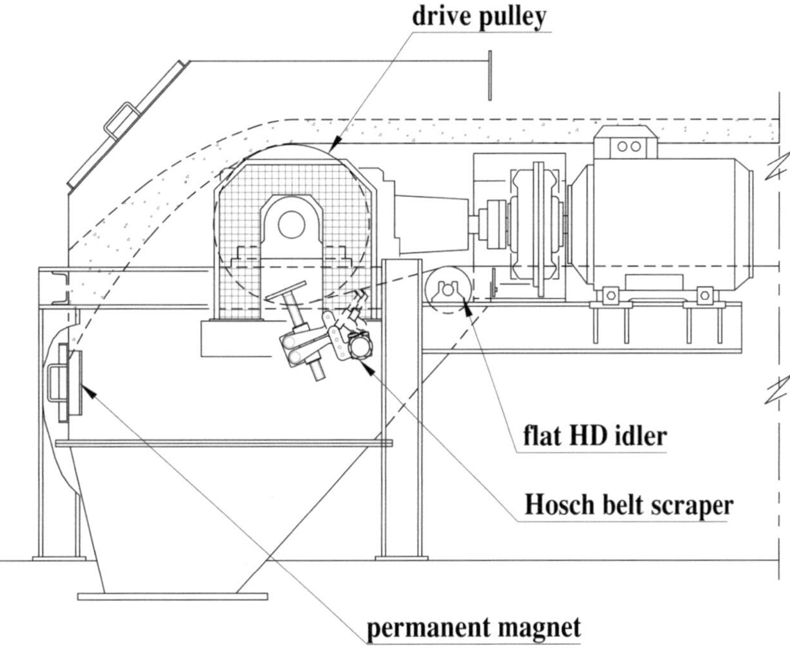

Figure 30.3. Example of design: conventional drive unit of a belt conveyor with an HD flat idler installed within the hopper instead of an outer snub pulley.

2.1.11 Magnetic Separators

To maintain the quality and the purity of the exported/processed bulk material and to protect downstream equipment from breakdowns, it is necessary to pull out the iron objects that accidentally fell into the bulk material.

There is a large number of different magnetic separators for pulling iron objects out of the material transported on the belt conveyor. For example, inline, suspended magnetic belt, suspended magnetic cross belt, stationary cross belt magnetic sheet, magnetic drive pulley, so on, are manufactured by Martin Engineering and Dings Magnetic Group in the USA; Eriez, UK and the USA; Electro Flux, India and by other manufacturers. Schemes of typical magnetic separators are shown in Figure 31.

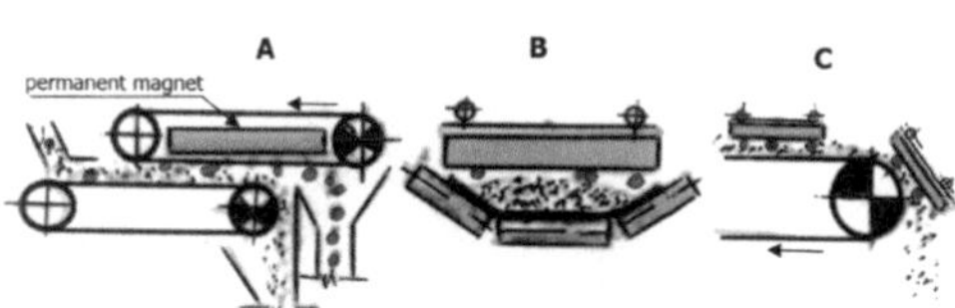

Figure 31. Typical magnetic separators (A is the inline magnetic belt, B is the suspended cross belt permanent magnet, and C are the suspended inline magnets).

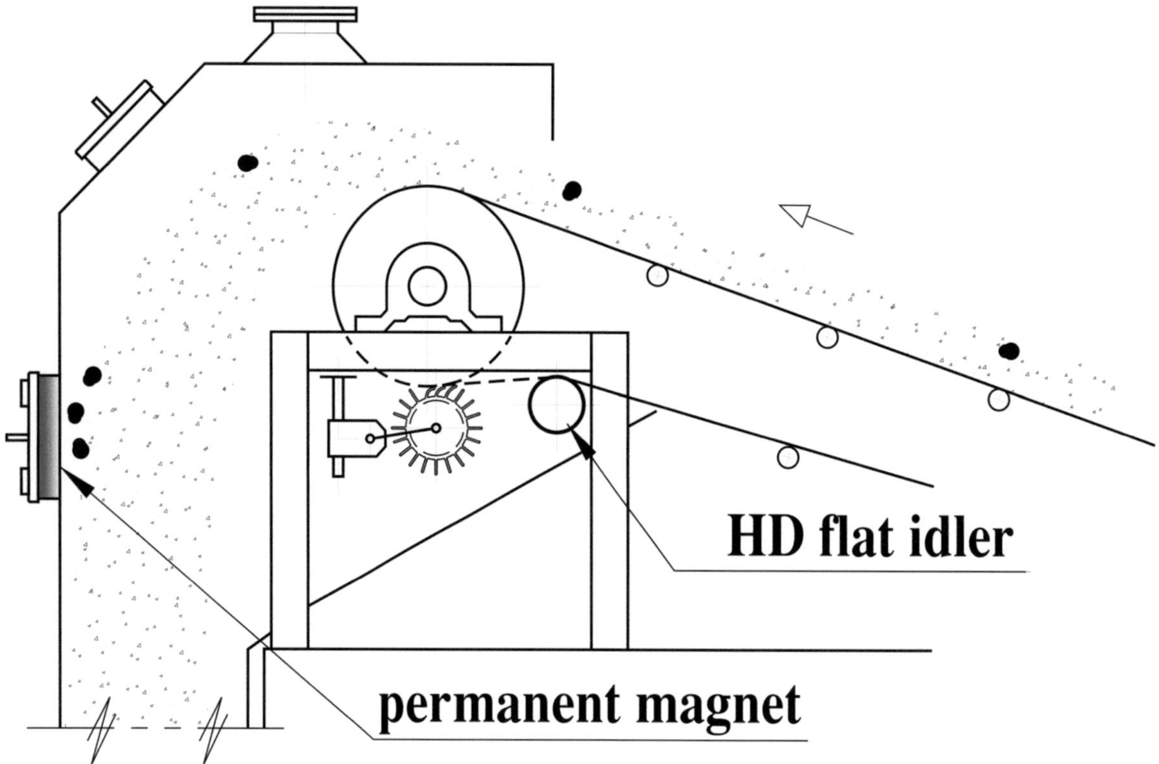

Figure 32. Permanent magnet installed on the manhole door of the head hopper.

As an option, the simple, inexpensive, maintenance-free iron-pull-out solution was implemented in the port of Eilat and the Tsefa facility. The flat permanent magnet (with a magnetic field of 14,000 Gauss), attached to the 316L stainless steel manhole door of the head hood, pulls out iron objects (up to 5 kg) from the flow of discharged bulk material (Figure 32). The maintenance team regularly opens the door to manually clean the magnet from the adhered iron objects.

The construction of the manhole door is shown in Figure 33, Figure 34, and Figure 35.

Figure 33. Manhole door of conveyor head hopper with installed permanent magnet.

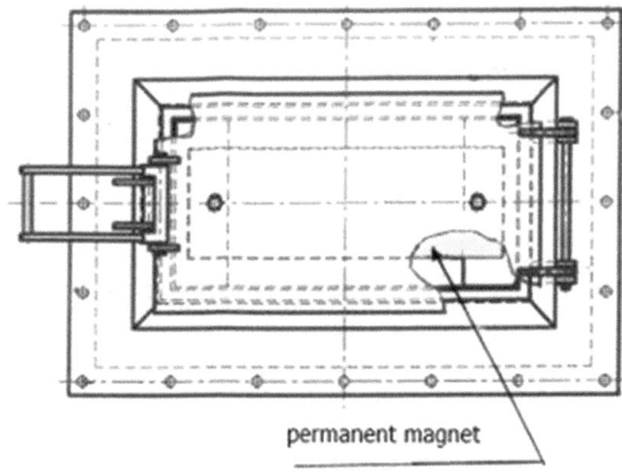

Figure 34. Manhole door with permanent magnet.

The anti-clockwise rotation of pivoted handle pushes downward and squeezes the gasket (made of soft rubber) helping the cam to pass on "dead point" and to lock the door (Figure 35).

Clockwise rotation of the handle squeezes the rubber gasket again and releases the cam to open the door.

This simple, reliable mechanism, developed by the author, rapidly opens and hermetically closes the door without bolts and nuts.

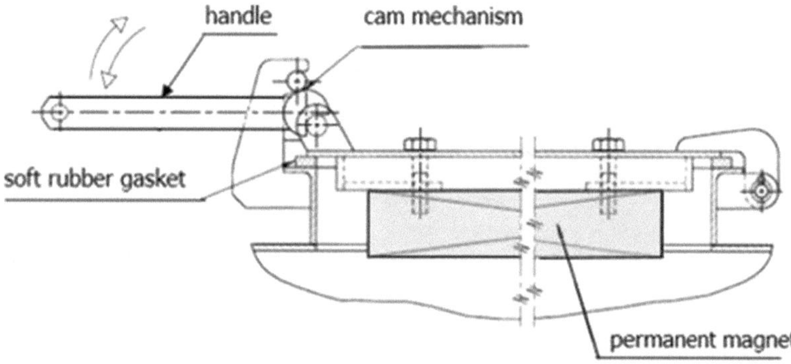

Figure 35. Cam manual opening/closing mechanism for manhole door (the door is closed) (see also Figure 33, Figure 34).

2.1.12 Diverters and Gates

Diverters are devices are used to change the direction of a flow of bulk material. Countless number of different types and modifications of mechanical diverters are in operation today.

Gravity diverters (Figure 36, A and B) and double-slide diverters (Figure 37) are used everywhere., and they have advantages and disadvantages.

Advantages:

Simple, compact, non-expensive diverters.

Disadvantages:

1. During material flow in the open-to-flow direction, dust is accumulated in and penetrates through seals in opposite-to-flow direction.
2. Abrasive bulk materials and rusty walls damage the rubber seals during blade (flapper) movement.
3. Damaged rubber seals cause leaks through side and bottom surfaces.
4. Rubber seals are located within an enclosed housing, thus, the damaged rubber seals are difficult to discover and their replacement is very hard and time-consuming procedure.

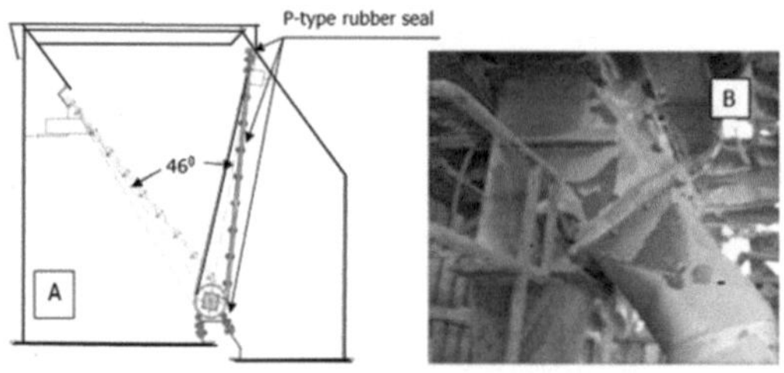

Figure 36. Example of gravity diverter.

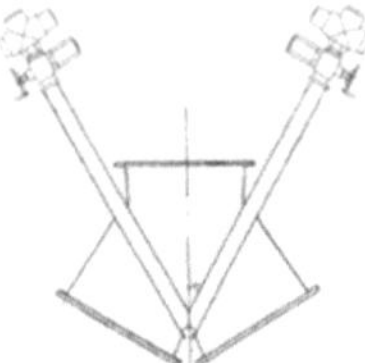

Figure 37. Dual-slide-gate diverter.

Our experience shows that for such bulk materials as potash, phosphate, apatite and so on, the right solution is the rotary (or rotating) diverter. Rotary diverters, that have been successfully operating since 1983 in the ports of Ashdod and Eilat, are reliable and, practically, maintenance-free devices. The principle of operation of rotary diverter is as follows:

- The rotating cone is suspended from outlet of the upper fixed hopper using slide bearings (15 mm thick and 15 mm wide Teflon rings) (Figure 38).
- The curved pipe is welded within the cone and connected the inlet of the cone to the opening in the bottom of the cone.
- The outlet pipes/chutes are fixed, supported by outer structures, and contact the bottom of the cone via rubber gaskets (P-type rubber is the best gasket for this application).
- The 0.5 kW gear-motor slowly rotates the cone, and when the centre line of the opening in the cone's bottom is opposite the centre line of the outlet pipe, the proximity sensor stops the rotation. Now material flow moves from the upper hopper via the curved pipe straight to the outlet pipe (Figure 38, Figure 40).
- Figure 38 shows that d_1, the diameter of the opening in the bottom of the cone, is less than d_2, the diameter of the inlet of the fixed outlet pipe, and this is done in order to compensate for a minor inaccuracy in the installation of the inlet and outlet pipes.

Two, three, or four outlet pipes can be in contact with the bottom of the cone as the cone rotation connects the hopper (through the curved pipe) in turn with every fixed outlet pipe and, concurrently, shuts all other outlet pipes/chutes.

Advantages of rotary diverters:

1. Reliable and practically maintenance-free diverters.
2. Two, three, or four fixed outlet pipes/chutes can be connected to the hopper in turn through an intermediate curved pipe located inside of the rotating cone.
3. There is no materials contamination: the cone's opening is automatically stopped opposite the required outlet pipe as other outlet pipes are completely separated from this opening.
4. The outer rubber gaskets of the outlet pipes are easy to inspect and to replace.
5. There is no dust emission during operation.
6. There are no large bearings sensitive to dust because the cone is supported by and rotates on narrow Teflon (or similar plastic) rings.
7. An additional outlet pipe can be fitted to the rotating diverter at any time without modification of the diverter.
8. The cost of the rotary diverter is lower than the cost of any three- or four-outlets diverter.

Disadvantage:

We found only one disadvantage of rotary diverter: its dimensions are much larger than dimensions of a conventional (gravity or slide) two-outlets diverter. However, this disadvantage becomes irrelevant if it is necessary to feed three or more outlet chutes from one inlet chute.

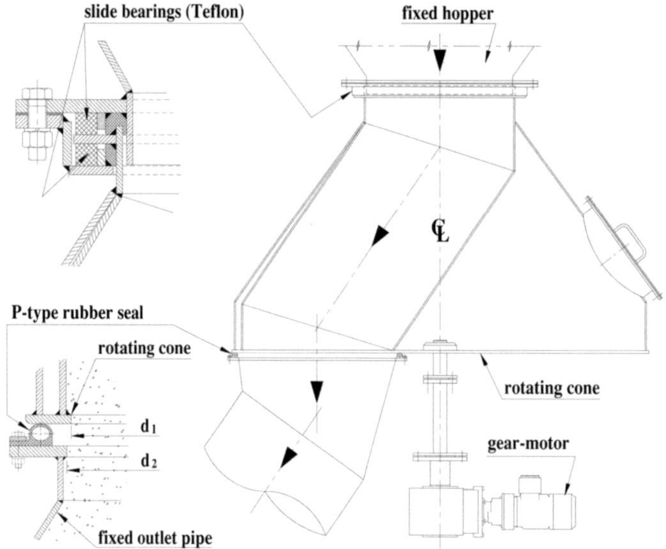

Figure 38. Principle scheme of a rotary diverter.

Figure 39. Operating rotary diverter.

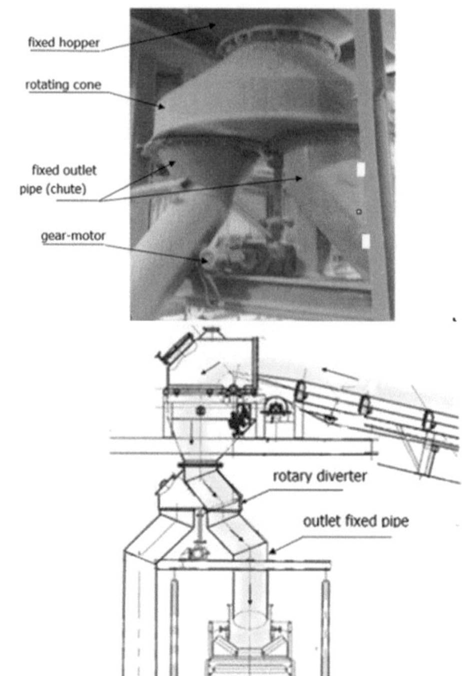

Figure 40. Example of rotary diverter installation.

When it is important to maintain a vacuum in the discharge silo, special double flapper valves (gravity actuated or motorized) are recommended.

2.1.13 Transfer Towers

The transfer towers (Figure 41, Figure 43, Figure 44) are high structural towers that are designed to enclose belt conveyor transfer points and thus prevent dust emission. Transfer point is the point where bulk material is poured from one belt conveyor to another conveyor in order to change the direction or to continue transportation.

A transfer tower inside of which are located conveyor drive units, tail sections, chutes, diverters, and so forth, are usually the main source of dust emission.

To reduce the dust, generated from falling bulk material from one belt conveyor to another inside the tower, insertable dust filters are widely and successfully used (Figure 42).

The DCE insertable dedusting device installed on the skirt boards of the tail section of the takeaway conveyor picks up and cleans about 75%÷85% of the dusty air.

Figure 41. The enclosed transfer tower. Four belt conveyor galleries come in and out of this tower.

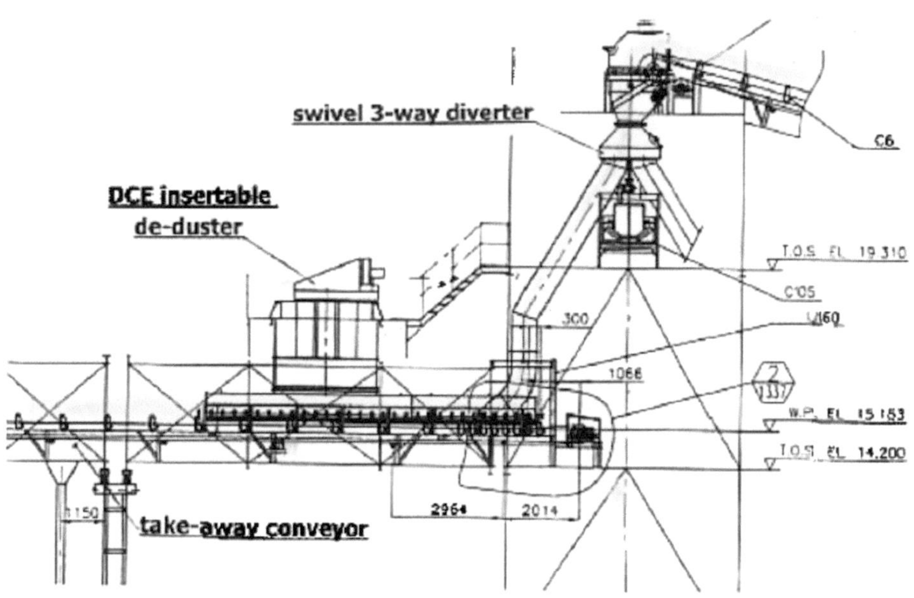

Figure 42. Typical transfer point (B1200 belt conveyors) equipped with insertable deduster.

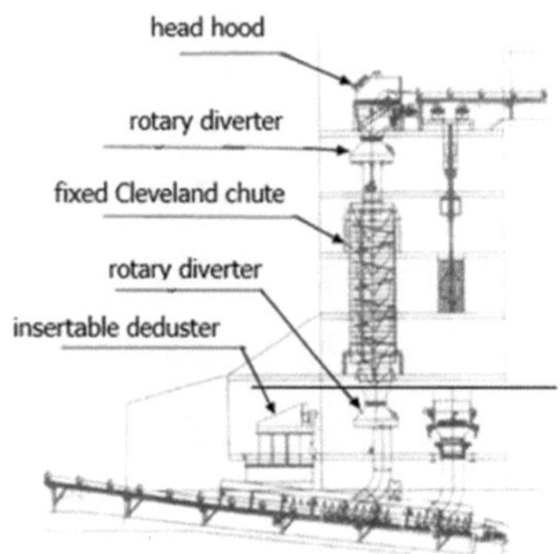

Figure 43. Transfer tower with fixed Cleveland Cascade chute and insertable deduster installed on the skirt boards of takeaway conveyor.

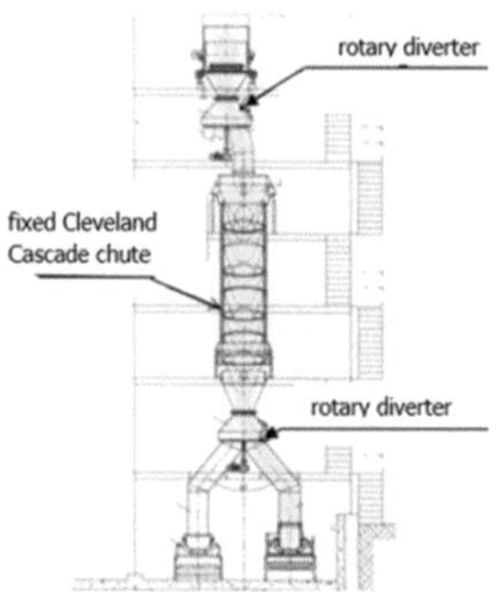

Figure 44. Transfer tower with two rotary diverters and a fixed Cleveland Cascade chute.

The transfer towers shown in Figure 43 and Figure 44 are designed with long vertical chutes (material drop height $\Delta H = 10\ m \div 20\ m$). The terminal velocity of the free fall of the bulk material reaches 15 m/s to 20 m/s, and the impact generates clouds of dust. The solution to this problem was found: the long vertical chutes were made up of a number of *fixed* Cleveland Cascade cones, zigzag assembled inside of the rigid frame. Such a chute (ended with a lower curved feed chute) reduces the speed of the flow of bulk material to 2.0 m/s ÷ 2.5 m/s, prevents impact of the bulk material and, hence, significantly reduces material degradation and dust generation.

2.1.14 Conveyor Belting

The belt of a conventional conveyor operates as a tractive element as well as a load-carrying element [12, 13, 14].÷

The calculated belt strength is limited by the reduced strength of the mechanical splices (by 20%÷30%) and by the decrease of the belt strength because of wear and aging.

The essential characteristics of a conveyor belt are as follows:

- flexibility
- transverse rigidity
- high tensile strength
- low elongation.

Specific requirements for the conveyor belt are as follows:

- oil resistance
- ozone resistance
- high temperature resistance
- wear resistance
- fire resistance.

There are two main groups of conveyor belts:

- abric multi-ply belts
- steel cord belts.

2.1.14.1 Fabric Multi-Ply Belts

The fabric belt is widely used today for industrial trough belt conveyors [12, 13].

The carcass of the belt consists of a number of plies of woven fabric or of solid woven fabric impregnated with rubber mix, and vulcanized together with top and bottom rubber layers (Figure 46).

The belt designation indicates the material used in both warp and weft.

Below are the different types of belts:

- EP (or PN in the USA), polyester (warp) and nylon (weft), ozone-resistant, oil-swelled changes its shape from flat to pipe-shaped.
- PP [or NN (nylon/nylon in the USA)] is less flexible, more elastic, and more
- sensitive to hot temperature.
- EE (or PP, polyester/polyester) in the USA) is an acid-resistant fabric belt.
- CC (cotton/cotton). Cotton gives good mechanical adhesion and high temperature protection.
- NBR (nitrile butadiene rubber) is heat- and oil-resistant, but the better the oil resistance, the worse flexibility. Also, this type is ozone sensitive.
- SBR (styrene butadiene rubber) is an oil-resistant belt.
- EPDM (ethylene propylene diene polyethylene) is a belt with good wear and ozone resistance belt, but it has a high degree of oil swelling. EPDM is resistant to most of the glues, so the cold splicing (cold gluing) of this type of belt cannot be usually carried out.
- PVC (polyvinyl chloride) is the solid-woven belt with top and bottom PVC covers (up to 4 mm thickness). The belt is widely used in mine conveyors.

Advantages of PVC belts: flame retardant (!), antistatic, oil-resistant, anticorrosive, resistant to impact damage.

Disadvantages of PVC belts: stiffness, more expensive (by 55%÷65%) in comparison with EP belt at the same rate.

- Today most industrial conventional belt conveyors use EP belting.

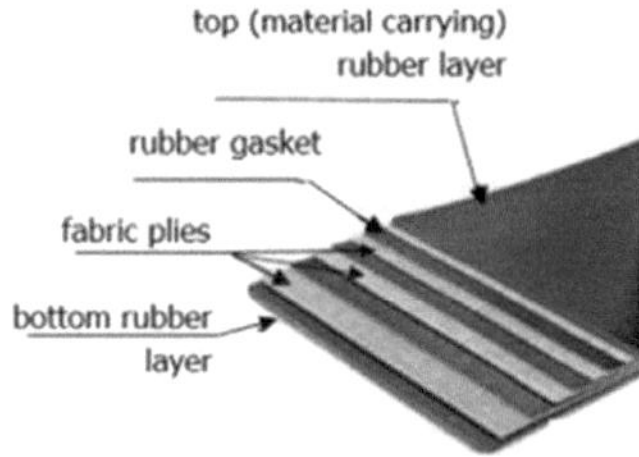

Figure 46. Typical fabric multi-ply belt.

The fabric multi-ply belts (Figure 46) are manufactured for various applications.

The strength of the fabric belt is determined by the number of fabric plies vulcanized together with the top, intermediate, and bottom rubber layers.

Often used industrial belt is EP with four plies (EP 630×4, 5×2). Five, six or more plies increase strength and, hence, permitted hauling tension of the belt, but this also increases the thickness and the stiffness of the belt, that can cause cracking of the rubber when the belt wraps over conveyor pulleys.

The acceptable hardness of the belt's top surface is from 65 to 75 Shore A.

An example of the standard technical data needed to fill out the purchase form to order an EP conveyor belt, is shown below.

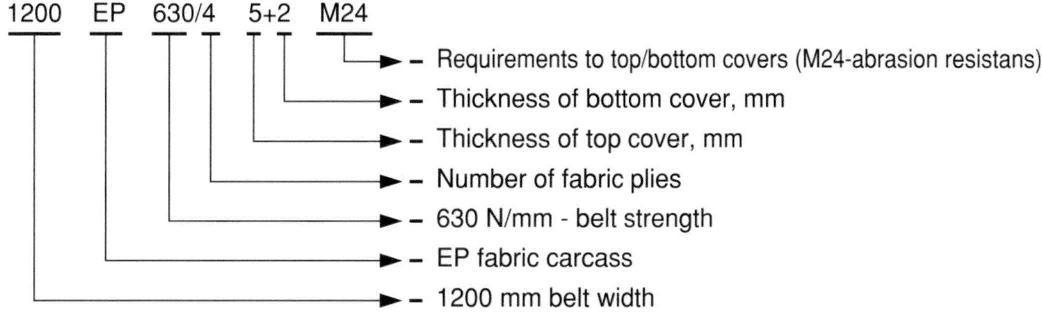

Special Conveyor Belts

- *Flame-Retardant (or Fire-Resistant) Belts (EN/ISO 340, EN 14973, EN 12882)*

According to ISO 340, flame-retardant belts must successfully pass the procedure test: the pieces of the belt (25 mm × 200 mm) are to be placed in a flame of +900˚C ± 100˚C for 45 seconds. After removal, the duration of the flame should be less than 45 seconds for each group of six tests (without the flame reappearing).

Flame-retardant conveyor belts are mainly used for internal transportation within buildings, in tunnels and in mines, where safety is of top importance.

- *Antistatic belts (EN/ISO 20284, ATEX).*

Antistatic belts are required in environments where coal dust, grain dust, or other dusty combustible materials are involved.

2.1.14.2 Steel Cord Belts (SCB)

Steel cord belts (SCB) are fabricated to increase hauling/pulling strength of rubber belts and, thus increase the length of the one-flight overland belt conveyors [14].

The conveyor hauling force is transmitted not through several fabric plies like this in fabric belts but through a big number of steel ropes moulded in one row into the rubber of the belt (Figure 47). For example, a SCB ST 2700 N/mm B1200 belt has 97 wire ropes, each 6.2 mm in diameter, moulded with a pitch of 11.2 mm.

Table 3

STEEL-CORD BELTING (according to ISO 15236, DIN 22131)

TYPE OF STEEL-CORD BELT	ST1000	ST2000	ST3150	ST5000
TOP LAYER THICKNESS, mm	4	4	6	8
WIRE ROPE DIA., mm	3.6	5.2	7.8	10.2
WIRE ROPE PITCH, mm	12	12	15	17
MIN. DRIVE PULLEY DIA., mm	600	800	1250	1600

Table 3 shows the specifications of a steel cord belt with a large number of steel ropes, diameter and pitch of which are determined by the required strength/rate of the steel cord belt.

Table 3 also indicates the minimum diameters of drive pulleys, recommended for SCB conveyors.

Figure 47 shows steel cord belts reinforced with additional layers of plastic mesh and/or wire grid(s) to prevent the belt from being ripped in a longitudinal direction by foreign sharp objects.

Figure 47. Typical construction of steel cord belts (ContiTech Group).

2.1.14.3 Fabric Multi-Ply Belts vs Steel Cord Belts

The typical crossover from fabric belt to steel cord belt is a belt tension of 1,000 N/mm.

Safety factor (SF):

Fabric belt SF = (8÷10): 1

SCB SF = (5.0÷5.5): 1

Minimum belt turnover length (TOL):

(TOL × B, where B is the width of the belt, m)

Fabric belt TOL = 10 × B

SCB TOL = 20 × B

Belt stretch (C/C or centre-to-centre, %):

Fabric belt E = 1.5% to 2.25%

SCB E = 0.25% to 0.3%

Pipe conveyor transition length (TL), (TL × D, where D is pipe dia., m):

Fabric belt TL = 25 × D

SCB TL = 50 × D

Transverse flexibility:

Fabric belt trough (angle of wing idlers) for B600 to B1500 is usually 30° to 35°.

SCB belt trough is usually from 35° to 45°.

Impact absorption at feeding conveyor section:

Fabric belt — great because of higher rubber content.

SCB — much less because of low rubber content.

Maximum Single-Flight (MSF) of a conveyor built up today:

Fabric belt MSF < 5,000 m

SCB MSF > 20,000 m

The current ratings (which may be increased in the future) of the maximum tensile strength of steel cord belts manufactured by the leading manufacturers are shown below:

Dunlop 5,400 N/mm

Metso 6,700 N/mm

Phoenix 8,500 N/mm

ContiTech AG 10,000 N/mm

For example, the rating of a steel cord belt at 600 N/mm break strength with a safety factor of 5, is 600 N/m × 5 = 3000 N/mm, or ST3000.

The steel cord belt's service life depends on the conveyor speed. The belt life is about 15 years at a speed of 4.1 m/s and only 10 years at a speed of 7.0 m/s.

The thickness of the top rubber layer of conveyor belt depends on the type of conveyed bulk material. Big, sharp lumps require thicker top layers to withstand heavy abrasion.

For ore transportation, the recommended thickness of the top layer of EP belt is 5 mm to 10 mm, as for granular to fine materials, the recommended thickness of the top layer is usually 4 mm to 5 mm. The thickness of the bottom layer is from 2 mm to 4 mm.

For the steel cord belt, the thickness of the top layer usually equals the thickness of bottom layer and varies (depending of rate of the belt) from 5 mm to 10 mm.

For horizontal overland conveyors, the quality of the belt rubber (belt roll resistance) and the tension resulting from the belt resistance to flexure, are among the main components of power consumption. For example, ContiTech AG was chosen to be the supplier of SCB for El Teniente's (Chile) new 20 km long four-flights conveyor. The belting was made from XXL compound, developed by the company and distinguished by improved viscoelastic properties. As a result, the belt roll resistance was reduced by 25%.

Kevlar belts utilize para-aramid fibres (instead of fabric plies or steel ropes) are used to increase the pulling strength of the belt. The fibres are sensitive to acids and salts and are very expensive.

2.1.14.4 Inclined Belt Conveyors

In accordance with IS 8730:1997 and CEMA [3], the recommended maximum angle of inclination of a belt conveyor directly depends on the parameters of conveyed bulk material. Minor changes in moisture content, a certain percentage of fine particles within granular bulk material, the shape and specific weight of the material particles, all these parameters significantly affect the choice of the maximum slope of the belt conveyor. Wet, non-free-flowing bulk materials can be transported at higher slope angles, while the easily fluidized powders can be transported at much lower slope angles.

Following are recommended inclination angles (RMI):

Bulk Material	**RMI**
Dry fine phosphate	11°÷13°
Dry potash (fine and granular)	14°÷15
Polished rice	8°÷ 9°
Dry gravel	15°÷17°
Coal	18°÷22°
Wood chips	25°÷27°
Ore	18°÷20°
Grain	13°÷14°

Below are a few simple practical recommendations allowing increase the slope of the existing belt conveyors:

- Reduce the conveyor speed by VFD (variable speed drive).
- Replace the belt with one wider than required and reduce the conveyor speed.
- Reduce take-up weight to the minimum required to prevent belt slippage on the drive pulley.
- Use a chevron-type belt.

Inclined conveyors with chevron belts (Fig. 48) allow increase of the conveyor slope angle by "replacing" low friction between smooth rubber belt and conveyed material with higher friction between material and material.

Advantages of chevron belts:

1. The slope angle of the conveyor can be increased by $2^{\circ} \div 5^{\circ}$.
2. New belts with fishbone profiles, moulded as an extension of the top rubber layer (e.g., Dunlop-Multiprofil), have a longer service life.

Disadvantages of chevron belts:

1. The service life of conventional projected chevrons is relatively short. They crack and break quickly by bi-directional wrapping over a number of conveyor pulleys.
2. Using a chevron belt on conveyors with gravity take-ups is not recommended because of fast damage of the profiles.
3. Regular belt cleaners/scrapers [15] are used for cleaning smooth belts only, so a chevron belt cannot be cleaned, resulting in significant spillage.
4. The cost of chevron belt is higher than the cost of the regular belt.

Fig. 48. Example of chevron belt conveyor (ContiTech Group).

2.1.15 Belt Cleaners

A thin layer of fine particles adheres to the smooth surface of the carrying side of the belt and, after discharge of the conveyor, gradually falls down. The spillage, which accumulates under the snub pulleys and return idlers along the conveyor route for some extended period of time, requires periodic cleaning which is a time-consuming and a labour-consuming operation.

To reduce the spillage, special belt cleaners are installed within lower hopper of belt conveyors. The scrubbers rub the return belt and drop the fine particles (or wet material) adhering to the belt back into the hopper [15].

We can find a large number of different belt cleaners on the market today, but we will consider only three types:

1. Sprung-blade scrapers, pressed against belt on the head pulley and used springier scrapers/blades to rub the belt (HOSCH, Rulmeca, Flexco)
2. Motorized rotating brushes (Rulmeca and others)
3. HR rotating dual-action brushes.

Sprung-Blade Scraper Cleaners

The principle of the scraper-type belt cleaner is shown in Figure 49.

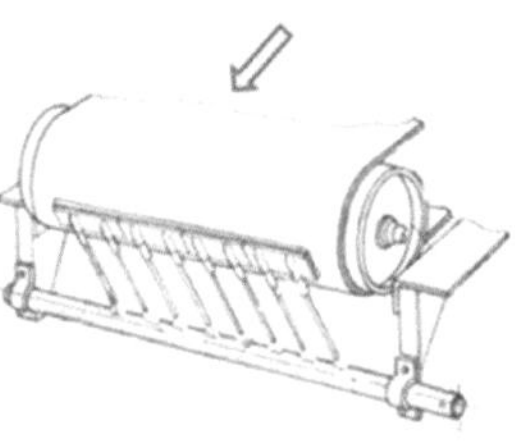

Figure 49. Work principle of scraper-type belt cleaner.

The Rulmeca scraper-type belt cleaners (Figure 49.1.) are used on conventional one-directional belt conveyors. The Flexco belt scraper-type cleaner is shown in Figure 49.2.

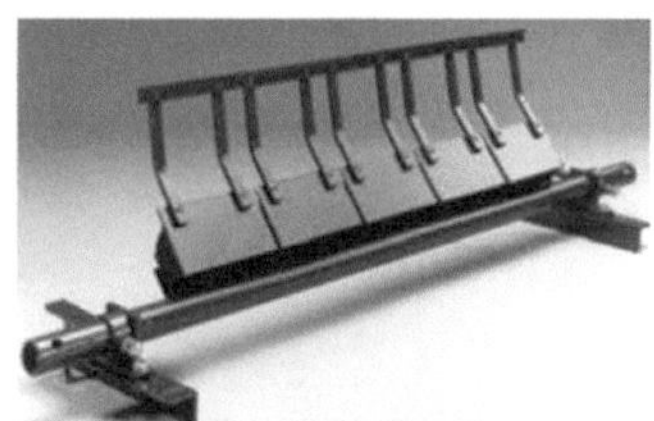

Figure 49.1. The Rulmeca scraper-type belt cleaner.

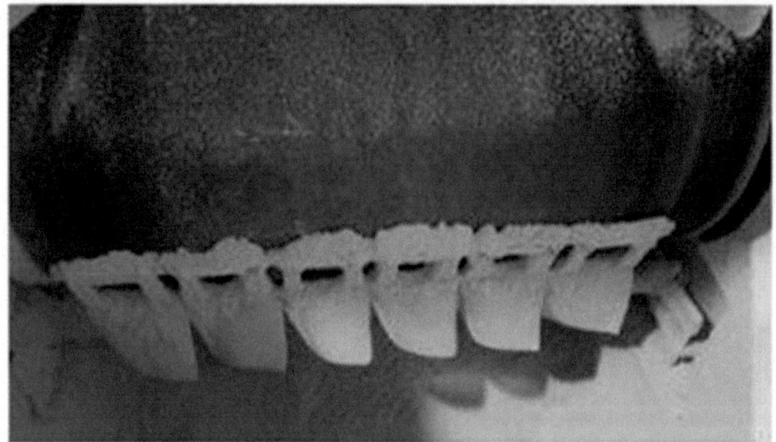

Figure 49.2. The Flexco (UK) HV2 scraper/cleaner.

Advantages of scraper-type cleaners:

1. Good cleaning efficiency, especially for wet, sticky bulk materials.
2. Can be used as primary and as secondary cleaners.
3. Suitable for the most difficult bulk materials.
4. No adjustment required until blades are worn.
5. Easy to install and easy to dismount.

Disadvantages:

1. Limited belt speed (< 2.5 m/s).
2. Problems when operated on reversible belt conveyor.
3. Problems when the springier blades interact with belt mechanical joints.

Motorized Rotating Brushes

The rotating brushes are installed under the head pulley and are in contact with the return belt. Rotating in the opposite direction to the belt movement, the brushes slightly rub the belt surface and throw the dislodged material back into hopper (Figure 49.3).

The brushes are mainly made of nylon or polypropylene.

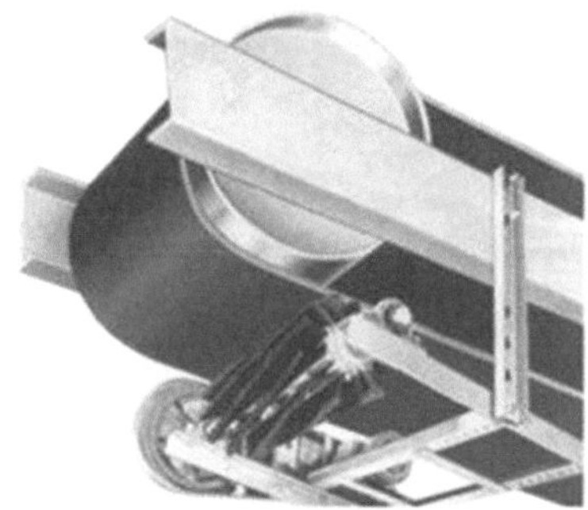

Figure 49.3. Belt-cleaning rotating brush (Rulmeca, Italy).

Advantages of rotating brushes:

1. Well clean the belt from dry, non-abrasive and non-adhesive materials.
2. Tension is automatically maintained.
3. Good for belts with mechanical joints.
4. There are no problems with cleaning reversible belt conveyors.

Disadvantages:

1. Limited operational temperature (up to +80°C according to Conveyor Components Co., USA).
2. Wear of brushes, rotating at high speed, should be visually checked regularly. Uneven wear causes imbalance and instability of the rotating brush, which leads to breakage of bearings.

HR Brushes

HR (hit and rub) rotating brushes were developed by the author and now work on two conveyors in port of Eilat.

This brush differs from conventional rotating brushes. Six wings, made from a four-ply fabric conveyor belt, are attached to the axis, which is rotated by a chain in opposite to the belt's direction. One chain sprocket is mounted on the shaft of existing snub or drive pulley, and the second sprocket is mounted on the axis of the brush. The sprockets are connected by endless chain to rotate the brush, so no special drive unit is required.

The brush is a double-action device: at first, wings hit belt and, continue to move, rub it and throw the removed material into lower the hopper (Figure 49.3).

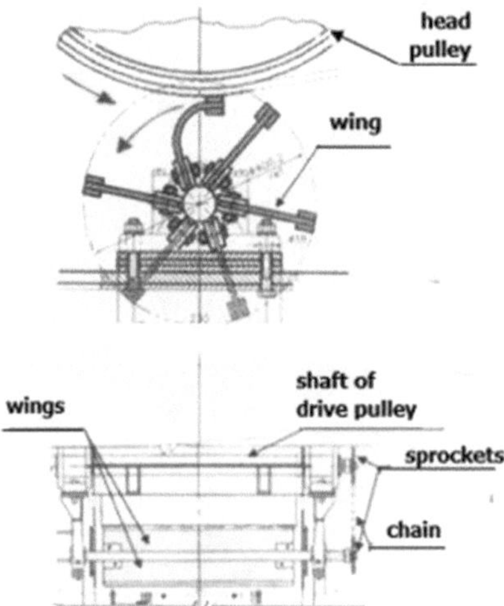

Figure 49.3. Operation of the HR rotating brush.

Advantages of HR brushes:

1. The double actions (hitting and rubbing) clean the belt throwing the removed material back into low hopper.
2. Suitable for high temperatures (up to +160°C).
3. Tungsten-carbide-tipped wings improve the wear resistance of the brush (tungsten outlasts the urethane by ten times).
4. Suitable for bi-directional belt conveyors.
5. Suitable for belts with mechanical joints.
6. Operate without a special drive unit, using rotating shaft of the snub (or drive) pulley to drive/rotate the brush.

Disadvantages:

1. HR rotating brushes are not manufactured industrially.
2. This is the tailor-made cleaning device, manufactured specially for each conveyor.
3. HR brush is not suitable for cleaning sticky, clay-shaped bulk materials.

2.1.16 Belt Conveyor Pulleys

The most used types of pulleys in use today are: smooth (Figure 50), rubber- or ceramic-coated (Figure 51), and vane-type pulleys (Figure 52) [16, 17].

Rubber- or ceramic-coated pulleys are most often used as drive pulley because they significantly increase "drive pulleys – conveyor belt" friction. Rubber coating is also used on bend pulleys to prevent/reduce adhesion of sticky bulk materials.

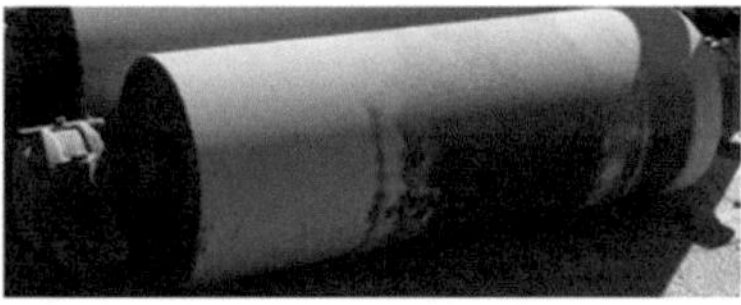

Figure 50. Smooth pulley.

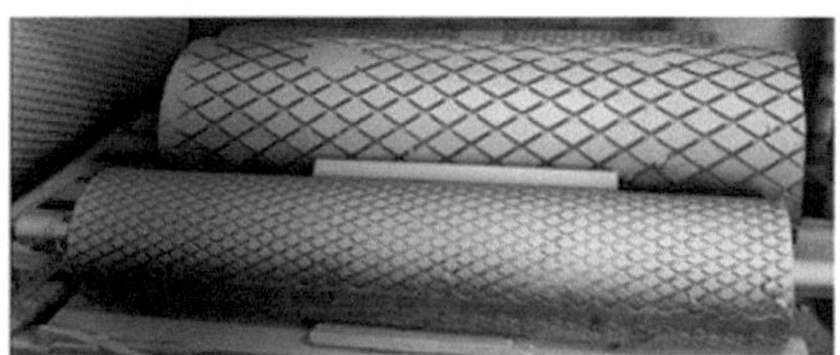

Figure 51. Rubber-coated pulley.

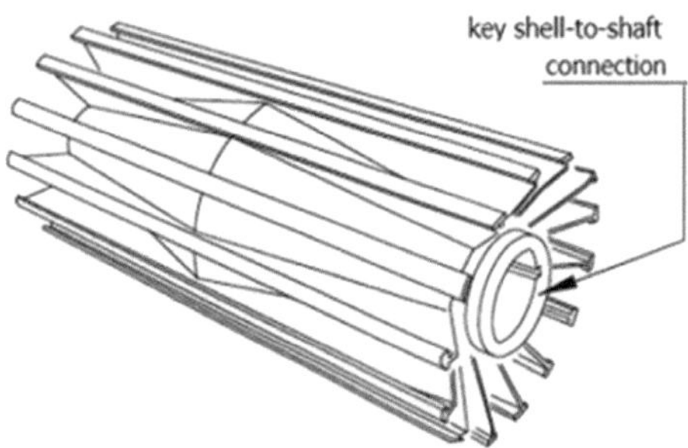

Figure 52. Vane-type pulley.

Diameter of the belt conveyor pulley should be chosen depending on the stiffness/thickness of the belt (or number of plies of a fabric belt) and on the ratio of the actual tension of the belt, running on the pulley, to the maximum allowed tension of the belt.

The experimental dependence: pulley diameter $D = K \times I$, where I = number of plies, and $K = 120 \div 150$ for conventional fabric belt.

Vane-type pulleys are mainly used as tail and take-up pulleys for conveyors transporting sticky bulk materials when residual material can be removed from the belt by light vibration (rapping) of the belt wrapping the pulley wings. The discharged bulk material slides down and away along the smooth decline cone surface of the pulley (Figure 52). The acceptable number of wings is twenty wings or more. A smaller number of wings (12÷16) causes strong vibrations since the belt jumps from one wing to the other, generating dust and spillage.

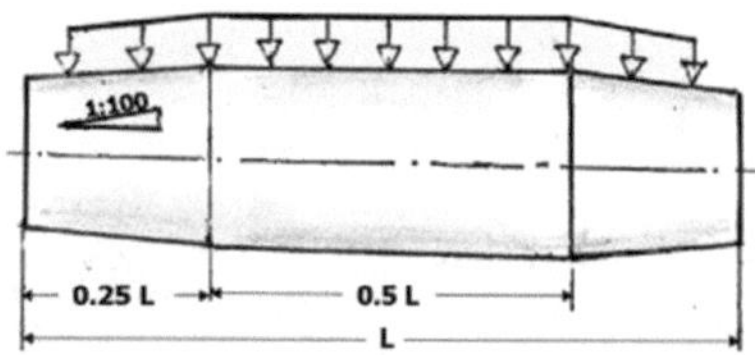

Figure 53. Belt tension distribution along typical belt conveyor bend pulley.

The rotation resistance of the pulley depends on belt tension, stiffness of the belt, angle of bending, bearing friction, and so on.

The shaft-to-pulley connections for bend, take-up, and tail pulleys are shown in Figure 54 and Figure 55.

There are two options for installing the pulley bearings:

a. external bearings
b. internal bearings.

Pulleys with external bearings are used in most operating today belt conveyors, whereas pulleys with internal bearings are used only in exceptional cases.

Disadvantages/advantages of pulleys with internal bearings vs pulleys with external bearings:

1. To replace a broken bearing, it is necessary to dismount the pulley and it's difficult and time-consuming operation, while the replacement of the external bearing is carried out in situ and without displacement of the pulley.
2. The regular checking and lubrication of the internal bearings, and measuring their temperature is the time-consuming operation: the conveyor must be stopped, the guarding must be dismounted and returned each time, while the lubrication and measuring the temperature of the external bearings can be performed on the running conveyor.
3. Considering conventional belt conveyors, we discovered no one advantage of pulleys with internal bearings over pulleys with external bearings.

Shaft-to-pulley connections.

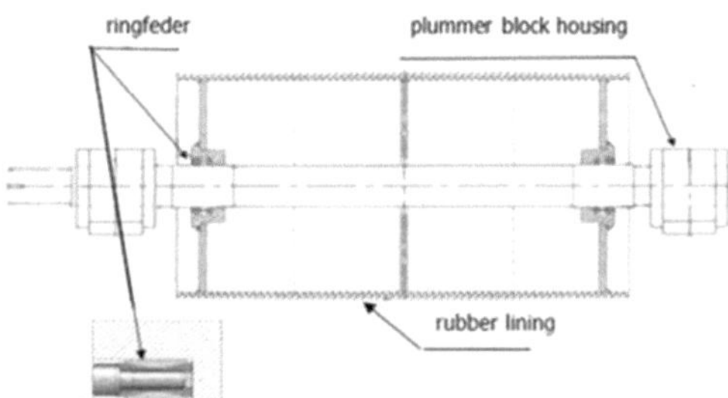

Figure 54. Ringfeder (RfN 7012) connecting the shaft with the pulley shell.

Today the Ringfeder (RfN 7012 or a similar connections) is the preferable pulley shaft-to-shell attachment, thanks to its reliability, simplicity, and ease of assembly and disassembly. Fig. 55 shows spare pulleys with welded bushings specially machined to install Ringfeders.

Figure 55. Spare pulleys with welded bushings machined to install Ringfeders for shaft-to-shell connection.

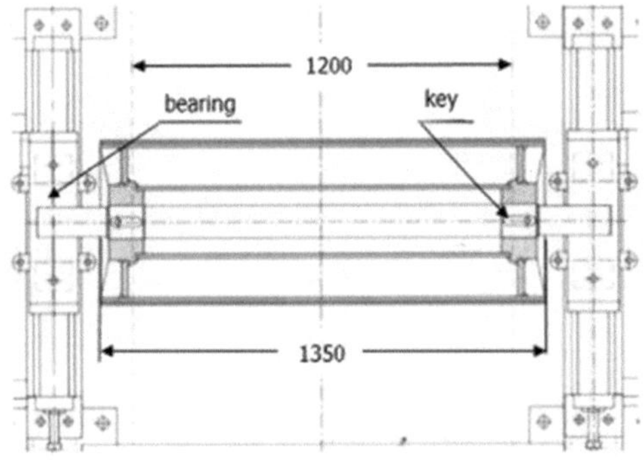

Figure 56. Key connected the shaft with a pulley shell (B1200).

The resistance of each snub, tail, and take-up pulley must be calculated (see below) and the resistance T_{e3} to be added to the sum of resistances for calculation of the effective belt tension T_e (see section 2.1.4).

<u>**Calculations of Conveyor Bend Pulley Resistance.**</u>

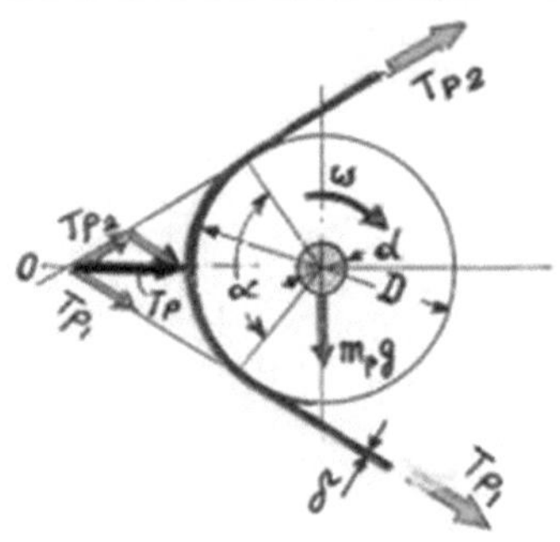

$$T_{p2} = z \times T_{p1},$$

where:

$$z_1 = [1 + f_R + f \times d/D \times \sin(\alpha/2)]/[(1 - \sin(\alpha/2) \times f \times d/D)]$$

where: $f_R = 1.23 \times \delta \times D^{-1.3}$ – belt stiffness coefficient, f-coefficient of friction of bearings (e.g., for taper roller bearings f=0.0018).

$$z = 1 + z_1$$

2.1.17 Idlers/Rollers

The most widely used types of carrying idler sets are trough offset three-idlers sets with a wing angle of 35° and idlers with a diameter 108 mm÷133 mm.

Trough carry and return idlers for conventional belt conveyors can be chosen in accordance with the recommendations [3, 18, 19, 20].

Conveyor Feeding Section

The feeding section is the most "sensitive" section of a conveyor where the conveyor is loaded by the flow of falling bulk material accompanied with impact and dust emission. In this section, the material *should be accelerated to the speed of movement of the belt* of the conveyor.

The optimal design of a conveyor's loading section must prevent impact of the flow of bulk material (especially big lumps) on the belt and prevent the belt from drifting to one side by forcing the material flow to run centrally.

In any case, the feeding chute must be smoothly curve-formed to prevent the direct impact of falling bulk material and in order to promote rapid acceleration of the material (Figure 42, Figure 43).

We will consider three options of feeding sections design:

1. Installation of impact-resistant idlers under the feeding chute.
2. Installation of impact bar section or combined impact bar–idlers section.
3. Installation of sets of standard idlers.

1.The *main disadvantage* of the impact idlers is the uneven wear of rubber discs causing imbalance and damage to idler bearings.

2. Special fixed ultrahigh molecular weight (UHMW) polyethylene bars (bed cradles) are often used instead of carry idlers (Figure 57).

The *disadvantage* of bed cradles is that the spilled bulk material can accumulate between the belt and the bars and stick to the rods, especially if the conveyor operates with long breaks. Adhesive material causes additional resistance to the belt movement and increased wear of the belt. For example, our attempt to use the bed cradles for the belt conveyor, transporting wet carnallite, failed – the carnallite "glued" the belt with the bars during a break, and it was impossible to re-start the conveyor.

3. In most cases, the optimal solution is the sets of the standard, inline, rubber-coated, 35° troughing idlers, fitted very close one to another (with the gap of 40 mm÷50 mm) under the loading chute,

Figure 57. Bed cradles (UHMW polyethylene bars) which are sometimes installed instead of idlers under the loading section of the conveyor (Rulmeca).

The spacing of the sets of carrying idlers along the conveyor route is determined by calculations of the permitted sag (2% or 3%) of the loaded belt between two neighbouring idler sets (see section 2.1.4). Trough carry idlers and return idlers for conventional belt conveyors can be chosen from [18].

Today, new conventional trough belt conveyors are mostly designed with the offset idler sets (Figure 58) instead of the traditional inline sets (Figure 59) or garland sets (Figure 60).

For example, offset idlers were installed on 19.1 km one-flight (with one booster) overland trough SCB conveyor commissioned in 2005 (Overland Conveyor Co., Luminant Mining, Texas, USA).

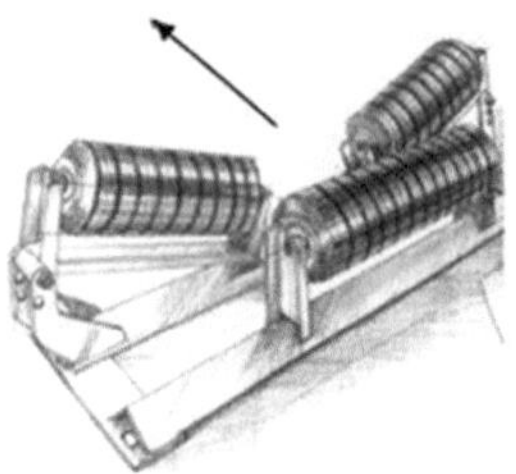

Figure 58. Typical set of offset idlers (Rulmeca, Italy).

Figure 59. Typical set of "inline" idlers (Rulmeca, Italy).

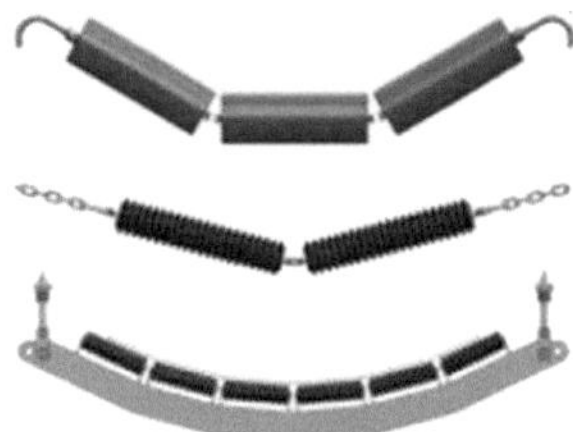

Figure 60. Sets of garland idlers (Rulmeca, Italy).

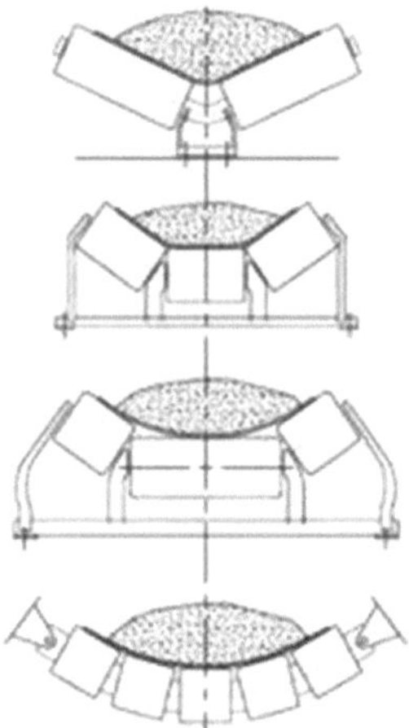

Figure 61. Cross sections of different belt conveyors.

Figure 62. Sets of garland idlers fitted to a rigid frame (Rulmeca, Italy).

Advantages of offset idlers:

1. Reduce material/dust clogging between central and wing rollers.
2. Increase supporting belt surface.
3. Prevention the belt tear as a result of roller dislodgment.
4. Eliminate idler junctions (how can this happen with inline idler sets, Figure 63).
5. Improved roller action against failure caused by roller failure "zipper" effect.
6. Much easier roller replacement, especially replacement of central idlers, which carry 60%÷70% of the full load applied to the idler sets.

Figure 63. The typical problem of "inline idler set – conveyor belt" junction.

The manufacturer's information (e.g., Rex diagrams [20] and others) will help the engineer to choose the optimal diameter of the idler and to define the shaft allowed load rating, the bearing load rating, etc., for any given belt conveyor.

The service life of an idler depends on the load applied, the idler's rotational speed, the environmental conditions, and the construction and reliability of the sealing. CEMA L_{10} life-span ratings are based on an idler rotational speed of $n = 500$ rpm. The slower speed of $n = 250$ rpm doubles the idler's service life, and a faster speed of $n = 600$ rpm shortens the life span by about 15%.

Special Idlers

Belt-tracking rollers (Figure 64) are used for the belt alignment and installed on the return run of the belt. They often successfully replace the training idlers installed on the top run of the belt.

Figure 64. Belt-tracking roller (Rulmeca, Italy).

The idlers, operating in a very wet and/or dusty environment, require a special construction of seals (double-seals) and additional deflectors to protect idler bearings from penetration of water and dirt.

The typical construction of high-quality sealing of idler bearings is shown in Figure 64.1.

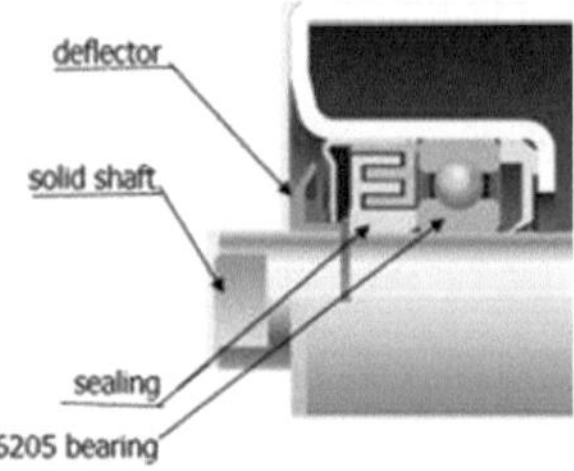

Figure 64.1. Construction of PSV2 idler sealing (Rulmeca, Italy).

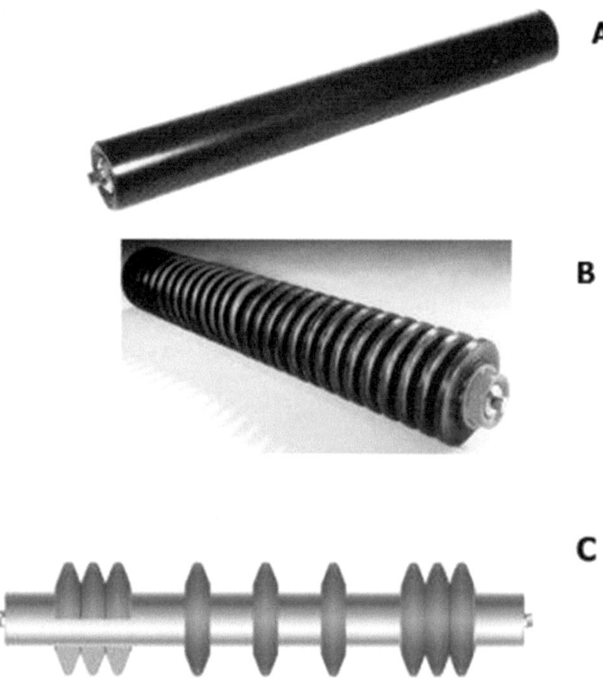

Figure 65. Different types of belt conveyor return idlers (Rulmeca, Italy).

The return idlers shown in Figure 65 are: (A) conventional flat smooth idler; (B) rubber spiral return idler with a great degree of belt cleaning and tracking; and (C) disc return idler (Rulmeca, Italy).

One of the new developments in idler manufacturing is the use of composite housings (super wear- and abrasion-resistant composite material idlers with double-labyrinth seal for carrying idlers, manufactured by Rulmeca, HSC Tech, and others). According to the HSC Tech company, the service life of these new composite idlers is two to three times longer than the service life of conventional idlers with steel housing. The additional advantage of composite idlers (according to the manufacturer) is their small rotational resistance (< 1.75 N).

Belt Conveyor Scales

Belt conveyor capacity can be controlled by belt conveyor scales (which usually use four load cells) with a degree of accuracy of 0.25%÷1%, or by weighing idlers (which much simpler, but have a lower degree of accuracy). Ramsey or Siemens weighing idlers operate at an accuracy of 1%÷2% for 60% to 100% of the belt conveyor's designed capacity.

2.1.18 Skirt Boards

To prevent spillage of bulk material on the conveyor's transfer/loading points, special guards (or skirt boards) are installed on the frame of a takeaway conveyor. The skirt boards are equipped with soft rubber strips (with a hardness ≤ 25 Shore A) that touch the conveyor belt and keep the loaded bulk material within the skirt boards (Figure 66).

It is not recommended to use conveyor belting as skirting strips. The exposed fabric plies trap fine particles and granules, resulting in severe abrasion and degradation of the conveyor belt top layer.

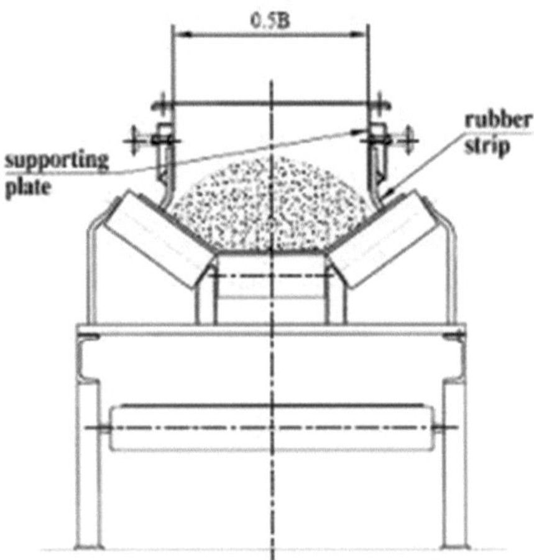

Figure 66. Principle scheme of skirt boards.

There are various types of skirt boards that differ from each other by fastening of the skirting rubber strip to the support steel plate. The problem is the choice of the right technical solution that is the simplest, most safe, allowing easy release and tightening of rubber strips in dusty and corrosive environment, and least expensive. Several types of skirt boards are shown below.

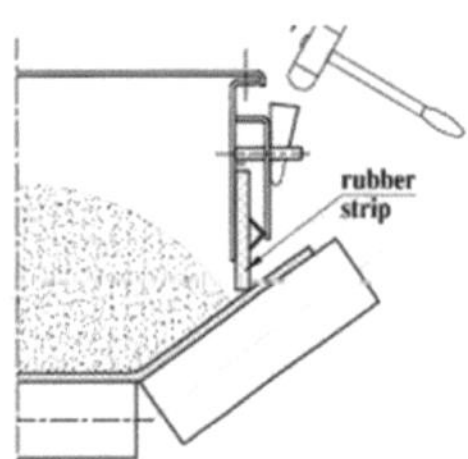

Figure 67. Method of skirt board rubber strip tightening, using a hammer

Advantages of the system shown in Figure 67:

1. Simplicity and reliability.

2. Low cost.

Disadvantage:

Use a hammer to lose and tight clamping components in a dusty and/or corrosive environment is a hard and unsafe work of a maintenance team.

The system shown in Figure 68 is the most common in material handling facilities today.

Advantage of the system:

The system is simple and reliable.

Disadvantage:

It is necessary to clean the nuts and then use an adjustable wrench to release the nuts and to lower or to replace the rubber strip in a dusty environment.

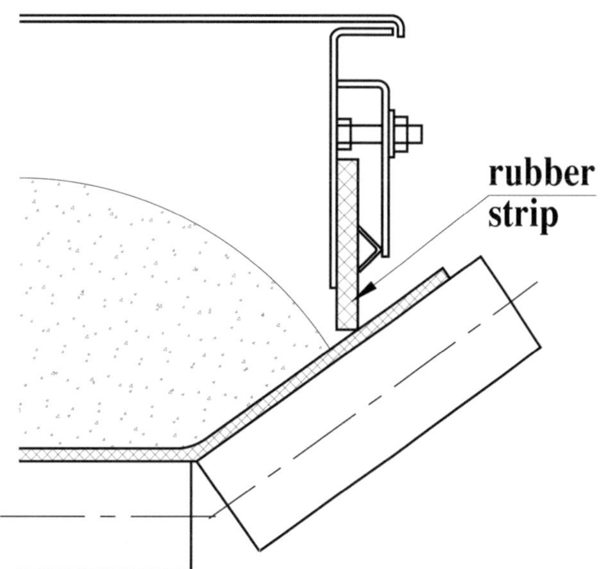

Figure 68. Skirt board rubber strip is connected to the supporting plate by bolts and nuts.

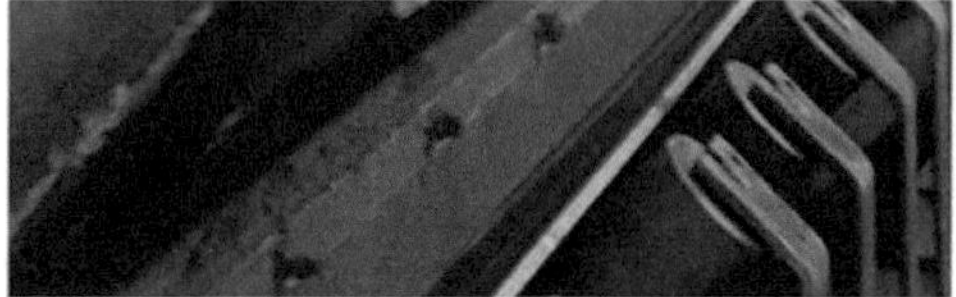

Figure 68.1. Skirt boards (see Fig. 68) on-site (rusty bolts and nuts).

The system shown in Figure 69, Figure 70, and Figure 71 is a simple and reliable skirt board solution that has been successfully used in ports for many years.

The system consists of a segmented steel skirt board (500 mm is the length of each segment) where the rubber strip is pressed against supporting plates by 50x50x5 angles via narrow metal sheets with welded 1/2″ bolts (pins) and hollow handles made from 3/4″ pipe (Figure 70).

A nut is welded to the end of the handle hollow pipe, so the threaded pin can go inside of the hollow handle by about 50 mm, while about 25 mm is enough to release the clamps and to lower or to replace the rubber strip without removing the handle.

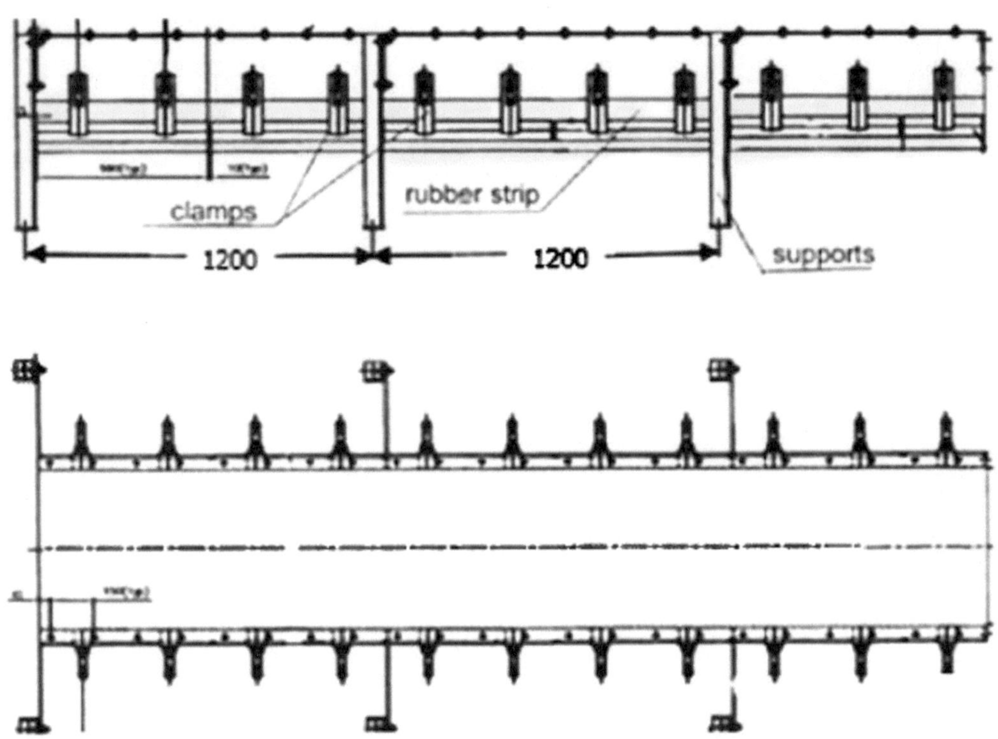

Figure 69. Skirt boards used in DSW (ICL) conveyors.

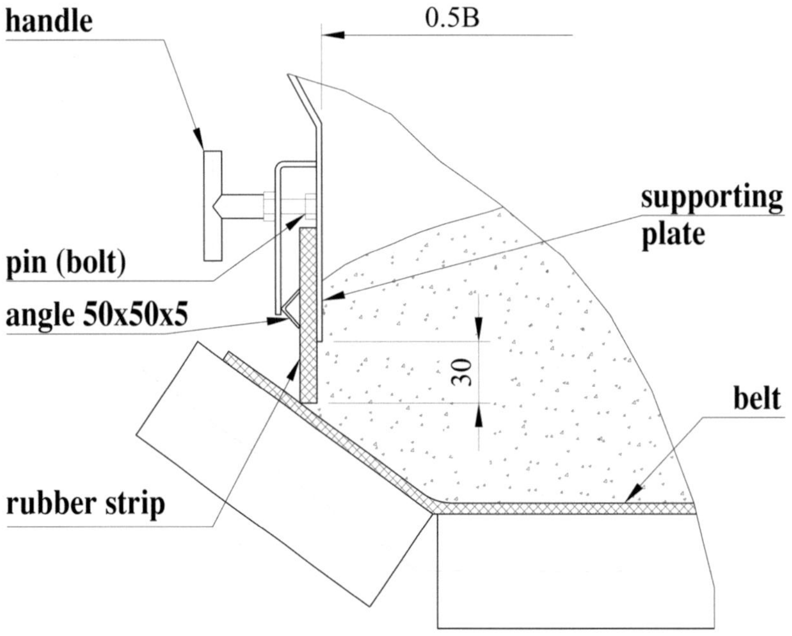

Figure 70. Cross section of skirt boards shown in Figure 69.

Figure 71. The skirt boards (Figure 69, Figure 70) installed on the operating belt conveyor.

Advantages of the skirt boards:

1. The release, the lowering or the tightening the rubber strips can be done without wrench or hammer, only by manually rotation of a T-shape handle.

2. Simple, reliable, safe, and inexpensive solution.

***Important Notes*:**

Below you will find six practical recommendations for how to prevent spillage and reduce dusting at the belt conveyor transfer point:

1. The optimal distance between skirt boards (B_1) should be 0.5 B÷0.55 B (Figure 69, Figure 70), where B is the width of the belt in m.

2. The tail section of takeaway conveyor: the recommended angle of rubber-coated throughing carry idlers is 35°. The idler sets should be installed very close one to another, with the gap of 40 mm÷50 mm.

3. The internal steel plate of skirt boards should be vertical and smooth as the gap between the plate and the conveyor belt is about 30 mm

4. The skirt strip should be made from soft rubber (≤ 25 Shore A), one piece per side.

5. For dusty bulk materials, is recommended to install on the skirt boards of the takeaway conveyor, the insertable deduster. Suction capacity of the deduster depends on capacity of the conveyor and on the height of the material falling [4]. For example, for 1200 tph conveyor capacity of dusty bulk material and height of material falling of about 2.5 m, the recommended suction capacity of deduster is 3000 $Nm^3/hr \div 3{,}500$ Nm^3/hr.

6. The feeding chute, entering skirt boards, must be smoothly curved in the direction of the belt's movement (Figure 42, Figure 43).

These six conditions (plus replacement of an outer snub pulley for an internal HD flat idler) also prevent material from drifting to the edge of the belt lessen dust drastically reduce maintenance and cleaning expenses at the transfer points.

To remove lumps and foreign objects from a return belt, special plough is recommended to install on the return belt, close to the tail pulley (Figure 72):

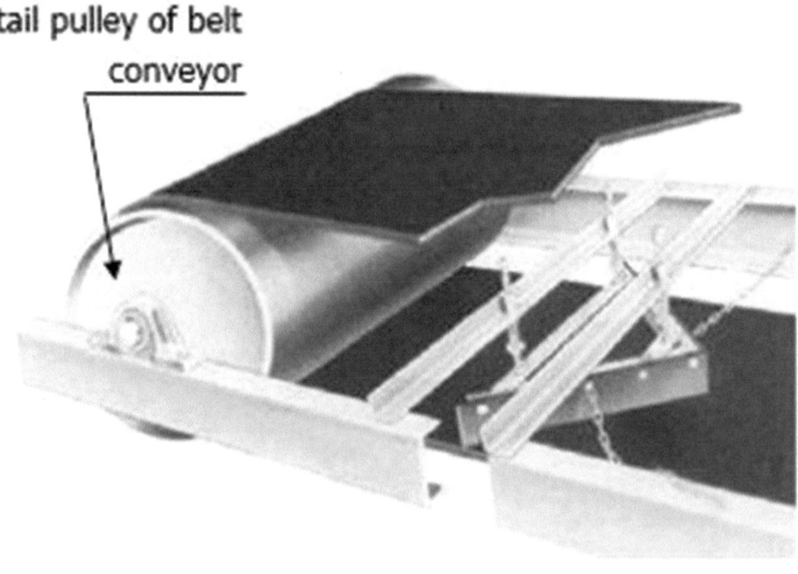

Figure 72. The special plough to be installed to clean return belt (Rulmeca, Italy).

2.1.19 Overland Conveyors

A generally accepted definition of an overland belt conveyor does not exist.

Belt conveyors that are 1.0 km long and longer, with a transporting capacity of 500 tph or more, can be identified as overland conveyors.

Each overland conveyor is the tailor-made system, designed to meet the specific conditions (conveyor route, number of curves, bulk material characteristics, capacity, ecological and climatic aspects, and so on) and requirements of the client (Opex, Capex).

In planning a conveyor route, extensive preliminary surveys must be carried out to determine the precise line of the route, and to work out how much earth will require to be moved.

The feasibility study must prove that the overland conveyor is more profitable and environmentally friendly than truck haulage.

The calculations of overland conveyors are much more complex than the calculations of conventional conveyors recommended by CEMA, DIN 22101, ISO 5048 and others. The calculations of conventional belt conveyors are based on the suggestion that the conveyor is a rigid system, while overland conveyors should be treated as flexible systems or a number of rigid sections connected by "springs", where stretch of the belt should be taken into consideration:

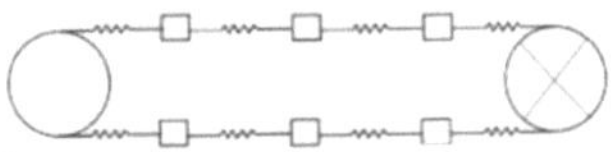

To accelerate and to stop the overland conveyors without belt tension waves, the conveyors should be calculated using the finite element method (FEM) [21].

We can find different types of overland conveyors that operate today in different countries of the world. We will consider six widely used conveying systems:

- overland conventional trough steel cord belt conveyors (SCB).
- overland Cable Belt conveyors.
- overland pipe conveyors.
- overland Flying Belt suspended conveyors.
- overland RopeCon suspended conveyors.
- overland Ropeway suspended conveyors.

2.1.19.1 Overland Trough Steel Cord Belt Conveyors (SCB).

The principle of operation of overland steel cord belt conveyor doesn't differ from principle of operation of a conventional belt conveyor.

The system of the two equations noted above (see section 2.1.4) is also relevant here.

The hauling tension, required to move loaded conveyor, is transferred to the belt via the friction force arising between the drive pulley and the belt. The difference between the two conveyors is: the belt of an overland conveyor should be exceptionally strong to withstand the huge required pulling tension. The fabric multi-ply belt is usually not strong enough to maintain this high level of tension, so overland belt conveyors are changed-over from fabric to the steel cord belt (see section 2.1.14.2) running on three-idler sets or on garland idler sets to be configured into a troughed shape (Figure 73).

Most of the overland belt conveyors in operation today in the world today are trough SCB conveyors.

Figure 73. Trough SCB overland conveyor (TAKRAF Tenova).

The cost of the belt can represent up to 60% of an overland conveyor's capital cost, so the designers and manufacturers of overland conventional conveyors make efforts to reduce the maximum tension and, in this way, to lessen the required strength (and the cost) of the belt.

To reduce the maximum belt tension (T_1) and increase the length of one-flight conveyors, the number of intermediate drive units (boosters or TT drives) can be installed on a SCB overland conveyors.

Such conveyors, called "multi-driven conveyors", are still relatively uncommon.

There are two main types of multi-driven conveyors:

- belt-to-belt TT boosters
- tripper-type boosters.

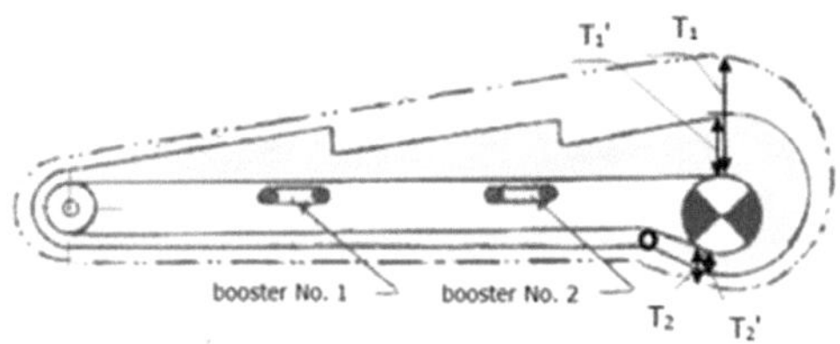

Figure 74. Belt-to-belt TT boosters reduce belt tension from T_1 to T_1'.

The belt-to-belt TT booster system consists in the short trough belt conveyors (boosters) that must be uniformly pressed against the bottom surface of the carry belt. The friction between two belts produces the additional intermediate hauling forces transferred from the booster to the main belt. These forces decrease the maximum tension force from T_1 to T'_1 and, respectively, reduce the requirements to the rating (strength) of the conveyor belt (Figure 76).

Advantages:

1. The system produces the additional intermediate hauling force(s) applied to the carry belt and thus lessens requirements for the strength/rating of the belt.
2. There is uninterrupted conveying, without intermediate transfer points.

Disadvantages:

1. The additional hauling force depends on the condition of the belts' surfaces (belt-to-belt friction coefficient).
2. The belts wear faster.
3. Additional drive units increase the Opex of the conveyor.
4. Higher Capex.

The tripper-type booster system (Figure 75) consists of several fixed trippers installed along the conveyor route, so the loaded belt "climbs" the tripper, wraps around the tripper head *drive* pulley (or around two drive pulleys), and drops bulk material back on the same belt conveyor.

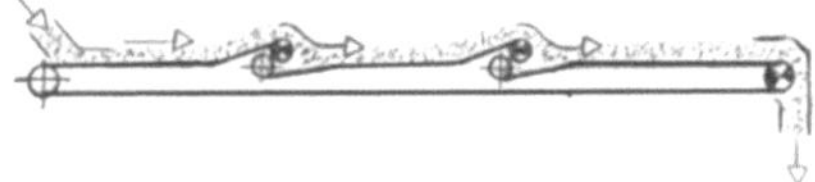

Figure 75. Overland conveyor with fixed drive tripper boosters.

Advantages:

1. The system creates additional intermediate hauling force which can be accurately calculated.
2. The additional hauling forces lessen T_1 tension and, therefore, reduce requirements for the strength/rating of the belt.

Disadvantages:

1. Bulk material, falling from tripper to belt, causes faster belt wear due to multiple accelerations of the conveyed bulk material.
2. Additional tripper boosters (drive units, chutes, skirts, belt cleaning, dedusting, and so on) increase Opex and Capex.

Horizontal Curves

Special, sophisticated calculations must be carried out to determine the balance of forces acting on the belt and on the material during the overland conveyors movement around horizontal curves.

The simplified scheme of these forces is shown in Figure 76:

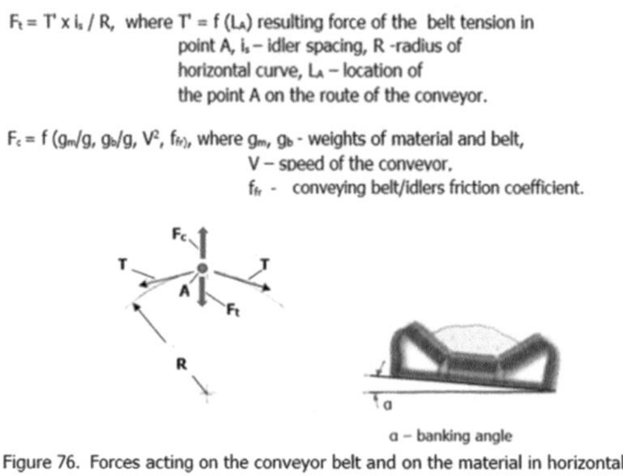

Figure 76. Forces acting on the conveyor belt and on the material in horizontal curves.

To prevent the belt buckling run-out from idlers on horizontal curves, the idlers should be banked (inclined) (Figure 76, Figure 77).

Figure 77. Horizontal curve of overland conveyor with gradual banking of the internal idlers (the belting was supplied by ContiTech Group).

It's possible to install overland trough SCB conveyors on the ground level with line stands mounted every 2 m to 7 m (Figure 78), or above ground level, at a height of 6 m to 40 m as the supporting piers/towers are located between 30 m to 50 m from one another (Figure 79).

Figure 78. Example of overland SCB conveyor installed on ground level (TAKRAF Tenova).

The longest overland conventional steel cord belt conveyor (ELB designed and built) is one-flight 26.7 km long with 17° inclined sections, which has a coal capacity of 2,400 tph. This conveyor, equipped with a B1200, St 2000 7/5 belt, was commissioned in October 2015 (Sasol Impumelelo, SA).

Figure 79. Example of above-ground SCB overland conveyor supported by piers (BEUMER Group).

In special cases, business may receive permission from environmental authorities to use open (uncovered) overland conveyors for the transport of specific bulk materials (Fig. 80).

For short conventional belt conveyors, the rollers resistance and flexure of belt and of material resistances are calculated according to average resistances recommended by CEMA [3],DIN [6], and these resistances are relatively small, but they become one of the decisive components in the calculations of horizontal overland conveyors where the power requirements are so high that all means should be taken to reduce the idlers' rolling and belt flexure resistances. It can be done using different means.

For example, the overland conveyor (BMH and CDI) 15.6 km long Zisco (Zimbabwe) using low-friction idlers and large-diameter central idlers (reduction of 0.75 N per idler set) with low energy loss on the belt's bottom layer (reduction of 11 N per idler set), was calculated according to a rolling resistance coefficient $f = 0.011$ instead of the DIN 22101–recommended coefficient $f = 0.0135$ and or the CEMA-recommended $f = 0.015$ [3]

Advantages of overland trough SCB conveyors:

1. Use of standard carry and return sets of idlers.
2. Relatively low rate of idlers replacement (usually 1% ÷2% of all installed idlers per year).
3. Use of a flat and smooth belt that can be cleaned within a discharge hopper by standard belt cleaners/scrapers (e.g. Martin, Rulmeca, HOSCH) to reduce spillage along the conveyor route.
4. New belts manufactured by ContiTech (Germany) with XXL compound have improved viscoelastic properties, minimizing rolling resistance by 25%.
5. Most of operating today overland belt conveyors are the trough SCB conveyors.

Disadvantages of overland trough belt conveyors:

1. The hauling of heavy belt tends to high energy consumption.
2. Splicing the belt by the hot vulcanization can be performed only by a high skilled team. This process requires that the conveyor be out of operation for one to two days.
3. Damage to the upper layer of the belt can allow atmospheric water to penetrate the belt, resulting in rust damage to the steel ropes.
4. Adjustment of a heavy, inertial belt (especially on the horizontal curves) is a complicated process reserved for a skilled maintenance team.
5. The reason why overland trough steel cord belt conveyors cannot break the Cable Belt record of 1982 for the world's longest one-flight belt conveyor (30.44 km long and with a capacity of 3,400 tph—Worsley Alumina, Australia) is that the strength of the CSB belt tension is limited to a certain number of steel ropes that can be moulded into the rubber along the belt's width. Inserting one row of ropes much larger in diameter, or installation of two layers of ropes in height, to increase the strength of the belt, will considerably increase the stiffness of the belt in the transverse direction. As result, it will be hard (may be impossible) to keep the trough shape of the belt. Another problem is the stiffness of the belt in the longitudinal direction: pulleys of extremely large diameters are required to wrap such stiff belt around.

The additional tail drives or multi-drive conveyors (Figure 74, Figure 75) can be used to reduce the required belt tension, as these systems significantly increase Opex and Capex of the overland conveying.

2.1.19.2 Overland Cable Belt Conveyors

The principle of Cable Belt - Metso conveyors differs from other conventional belt conveyors.

As noted above, in the conventional trough belt conveyors, two functions — the hauling the belt and the carrying of bulk material — are performed by one component: conveyor belt.

In the Cable Belt system, these two functions (hauling the belt and carrying the material) are separated.

The belt is designed for carrying bulk material only, as the hauling function is realized not by the belt but by the two endless steel ropes (Figure 80) and the belt, loaded with bulk material, only "rests" on the two ropes.

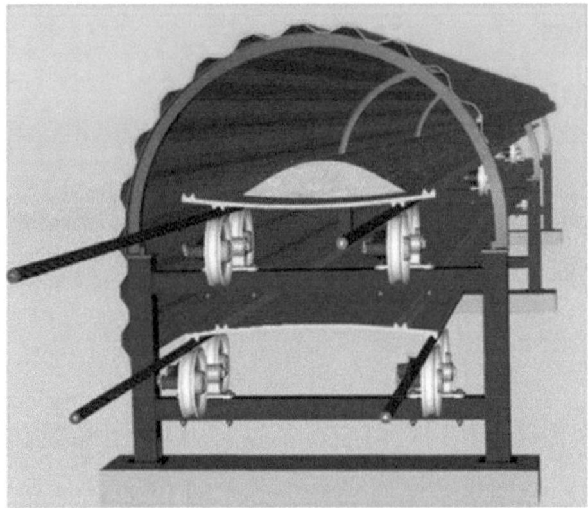

Figure 80. Principle scheme of a conveyor line stand (Cable Belt - Metso).

Figure 81. Picture of the line stand of the operating conveyor (Cable Belt – Metso).

The nominal pitch of line stands (constructions with installed poly-pulleys, supporting the ropes, Figure 80, Figure 81) of Sdom-Tsefa Cable Belt conveyor is 7 m. The pitch of stands supporting return run of ropes with empty belt, is 14 m. On "brows" (vertical curves) the pitch of line stands is reduced to 4 m to lessen the load acting on the poly-pulleys from inclined ropes.

Figure 82. Poly-pulleys support hauling wire rope (Cable Belt – Metso).

Figure 80 shows two endless ropes carrying the loaded top belt and the same two ropes carrying the empty bottom belt in the reverse direction.

The belt of the conveyors is manufactured with so-called "shoe forms" moulded on the belt top and bottom surfaces (Figure 83).

The cross section of each shoe form has the shape of a chevron (Λ) to grip the rope (Figure 87).

As the belt simply rests on the ropes, the belt is not tensioned (except for the minimal stretch caused by the main rope elongation/stretch and the belt take-up tension of about 3,000 N applied to the belt just passed head pulley).

The slightly tensioned belt can be easily mechanically spliced or replaced on the spot.

The main ropes safety factor of 5:1 was defined by the National Coal Board (UK) for Cable Belt conveyors operating in UK coal mines.

For the ropes of the Sdom–Tsefa overland conveyor, the safety factor of 3:1, chosen by the Cable Belt Co., UK, was approved by the DSW (ICL).

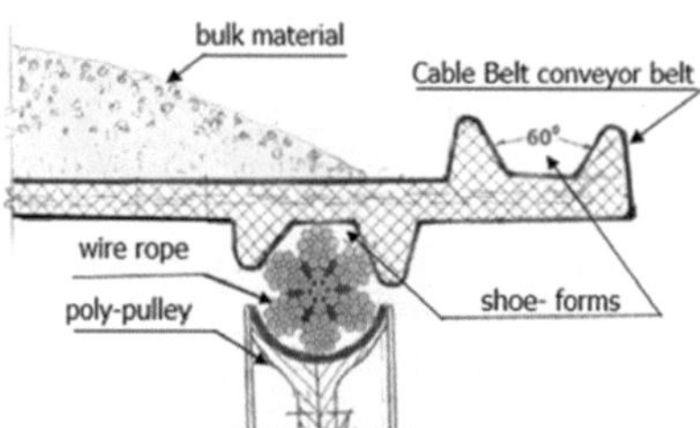

Figure 83. Cable Belt - Metso conveyor: interconnection of belt, wire rope, and poly-pulley.

The Cable Belt conveyor belts are manufactured with a pre-calculated transverse stiffness because the over-flexible belt can bend under the weight of the bulk material and fall between two supporting ropes, as the over-stiff loaded belt is running flat, resulting in side spillage of the bulk material along the conveyor route.

The belt deflection of Sdom-Tsefa Cable Belt conveyor, fully loaded with the material (47.5 kg/m) is about 70 mm. This deflection fully eliminates the side spillage of potash.

The pictures below show the Sdom–Tsefa Cable Belt overland conveyor, 18.2 km long with 800 m in lift, which has been operating since 1987 at a capacity of up to 650 tph, with a speed of 3.8 m/s and a specific power consumption of about 0.25 kW per km x ton/hr. The drive unit of the Sdom–Tsefa conveyor includes two 2,000 kW synchronous electric motors, a gear reducer with a mechanical differential, and two Koepe (drive) wheels 4 m in diameter providing (via friction between the ropes and the Koepe wheels lining) the hauling force to the main steel wire ropes (Figure 84).

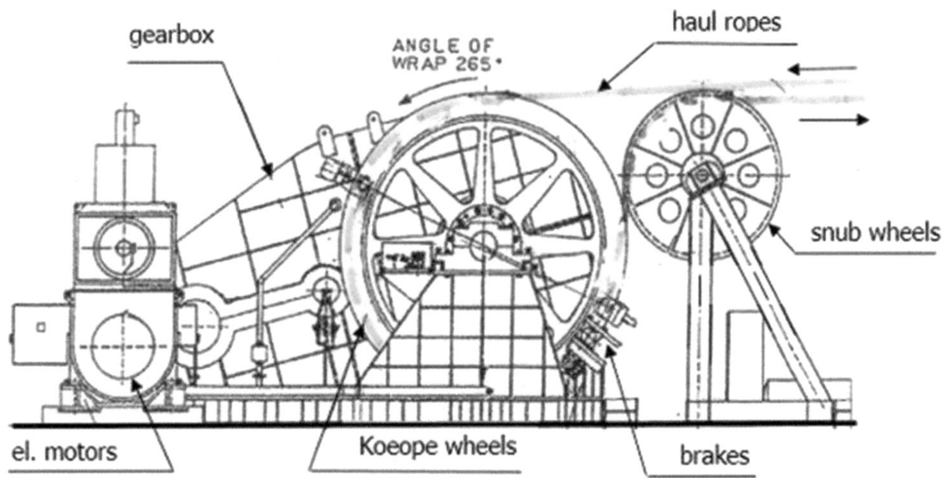

Figure 84. Drive unit of 18.2 km Sdom–Tsefa one-flight overland conveyor driven by two 2,000 kW synchronous electric motors (Cable Belt – Metso).

Figure 80, Figure 81 show typical line stands, consisting of a covered frame with attached pivoted rockers. Each rocker carries two poly-pulleys supporting the rope (Figure 82).

The take-up arrangement of the Tsefa Cable Belt conveyor is shown in Figure 85, and the take-up carriage for the main rope is shown in Figure 86.

The length of the tracks for the take-up carriages should be long enough (300 m ÷ 400 m) to gather the difference between the ropes maximum tensioned by tension T_1, and the ropes that looped between line stands by low, take up tensionT_2, and, in addition, to gather the elastic elongation of the ropes.

The elastic elongation of a rope can be calculated as follows:

$$\Delta L = W \times L / (E \times F),$$

where W is the applied load, kg; L is the rope length, m; E is the elastic modulus of the rope, (5,000 kg/mm² ÷ 7,200 kg/mm²); and F is the area of the rope, mm².

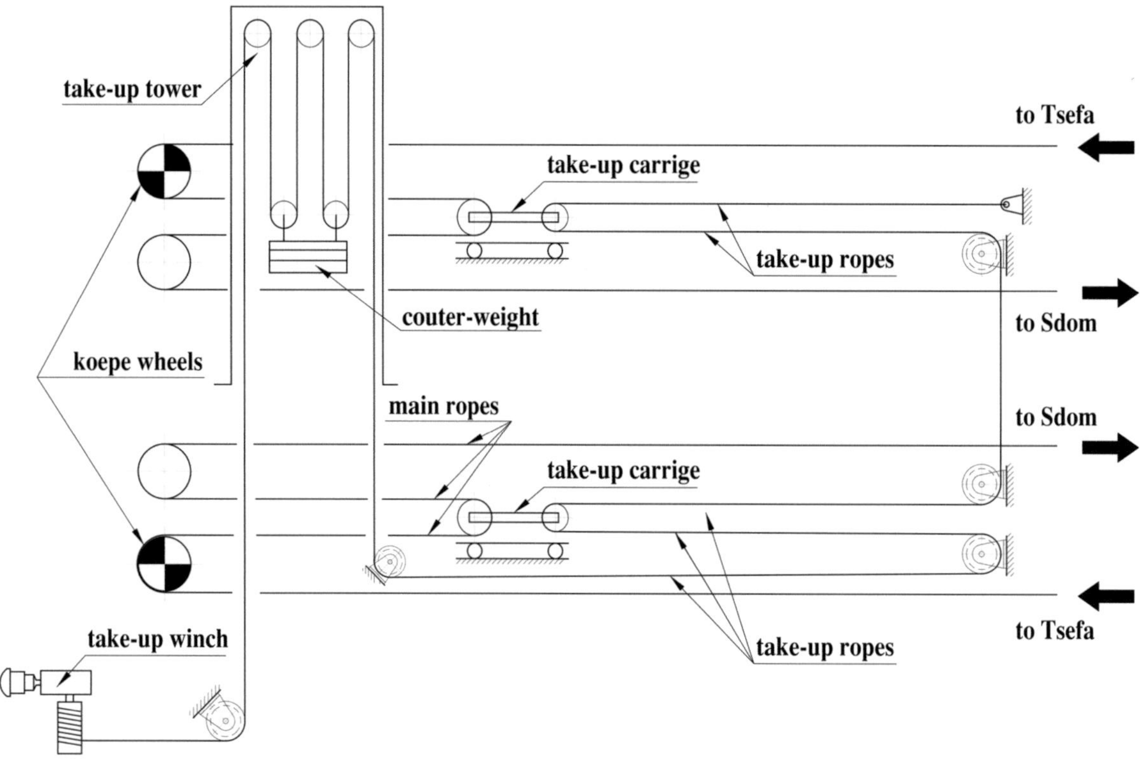

Figure 85. Principle scheme of Sdom–Tsefa Cable Belt conveyor gravity take-up system.

Figure 86. Take-up carriages of the main ropes of Cable Belt overland conveyor.

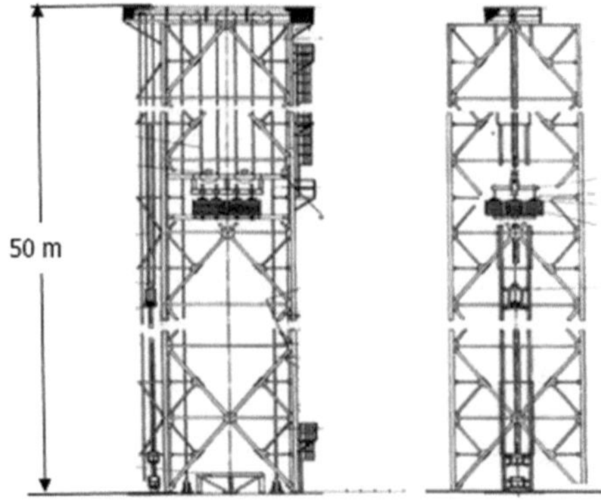

Figure 87. Take-up tower of the overland conveyor (Cable Belt -Metso).

Figure 88. A picture of the take-up tower.

The forces that arise during movement of the conveyor along horizontal curve, push the belt off the ropes. To prevent this type of failure, the Cable Belt-Metso uses one of two options:

- inclined line stands (angle of the inclination is between 2° and 5°) (Figure 89)
- inclined poly-pulleys (Figure 90).

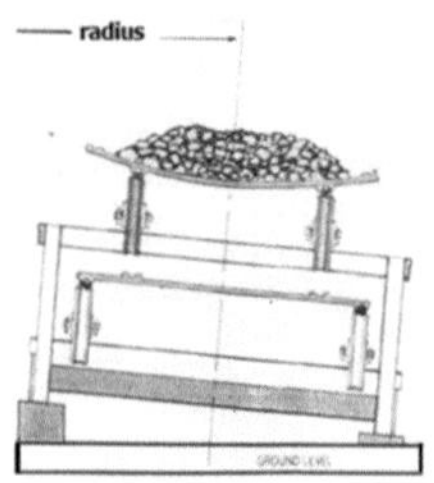

Figure 89. Inclined line stand on horizontal curve (Cable Belt -Metso).

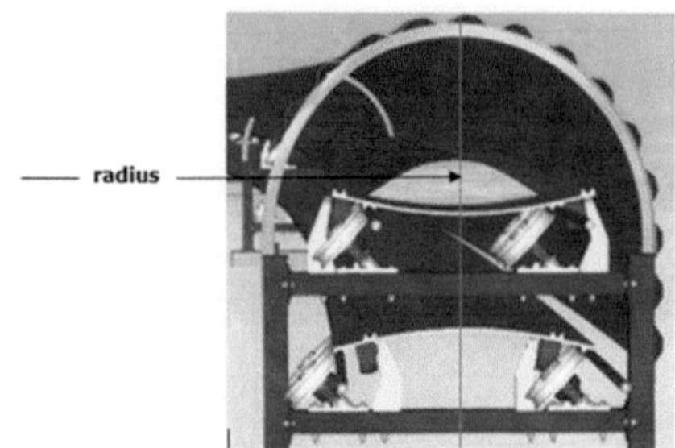

Figure 90. Inclined poly-pulleys on sharp horizontal curve (Cable Belt-Metso).

Inclined poly-pulleys allow the Cable Belt conveying route to have sharp horizontal curves as route of the Line Creek conveyor (Figure 91).

Figure 91. The Line Creek conveyor (Cable Belt - Metso).

Figure 92 shows as the Sdom–Tsefa Cable Belt conveyor is "climbing" at a height of about 800 m with a slope angle of about 18°.

The longest Cable Belt conveyor in the world, a 30.44 km one-flight conveyor operating since 1982 is shown in Figure 93 (Worsley Alumina, Australia).

Figure 92. The 18.2 km Sdom–Tsefa one-flight Cable Belt overland conveyor "climbing" the height of 800 m at an angle of 18°. The view from Sdom to Tsefa, Israel.

Figure 93. The longest in the world, a 30.44 km one-flight Cable Belt conveyor at Worsley Alumina, Australia (Metso).

Figure 94. Typical poly-pulley used for Cable Belt conveyors.

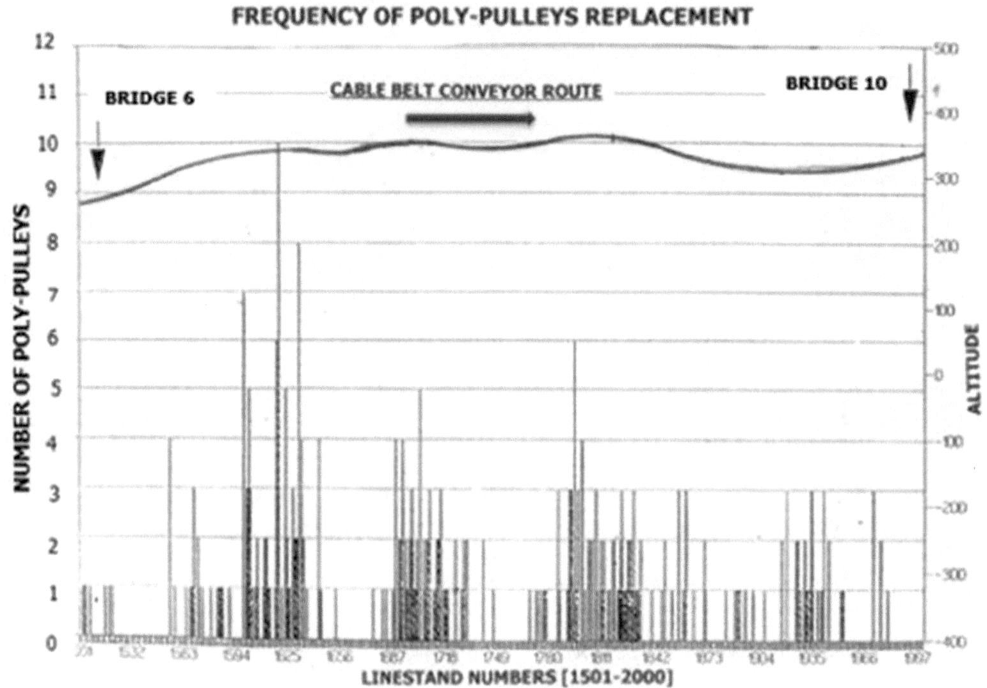

Figure 95. Graphic presentation of frequency of poly-pulley replacements.

Advantages of Cable Belt conveyors:

1. Reliable conveying, as the belt simply "rests" on the ropes (using Λ-type shoe forms, Figure 83), so the belt is guided by the ropes. No adjusting of the belt is ever required. Even non-central feeding of the belt doesn't cause the belt to dislodge or run out from the ropes.
2. The conveyor hauling tension is transferred via wire ropes and not via the belt, so the belt is practically non-tensioned.
3. The relatively light, non-tensioned belt can be easily handled (mechanically spliced, re-spliced, shortened, lengthened, partially replaced, and so on).
4. The service life of the belt at normal operation can reach 15÷20 years.
5. The inclination of line stands or the inclination of poly-pulleys within vertical line stands (Figure 89, Figure 90) allows realize conveying routes with small radii of horizontal curves.
6. From 1982 the Cable Belt conveyors hold the world record of the longest, one-flight belt conveyor (31.44 km, Worsley, Australia).

Disadvantages of Cable Belt conveyors:

1. The poly-pulleys should be replaced at yearly rate that can reach 10%÷20% of the total number of installed poly-pulleys (Figure 95). These frequent replacements require daily patrolling along the running conveyor and detecting the worn poly-pulleys by listening for the sharp noise that occurs once the poly-pulley lining has been worn down by the wire rope (for comparison, SCB conveyors require weekly patrolling along the conveyors).
2. The poly-pulleys can be replaced only when the conveyor is stopped. After replacement of the polyurethane lining, most poly-pulleys are sent back to a return service. An example of a poly-pulley replacements report is shown in Figure 95. Metso Co. has been constantly seeking to prolong the service life of poly-pulleys.
3. The Cable Belt conveyor belt is the tailor-made belt, designed and manufactured specially for the particular overland conveyor.

2.1.19.3 Overland Pipe Belt Conveyors

Pipe conveyors represent a relatively new technological direction in the development of overland belt conveying.

The loading, unloading, and take-up sections of the pipe conveyor are the same as those of any conventional trough belt conveyor, but, after the belt is loaded, the special installed rollers are gradually rolled up the belt to make an overlapped circle/pipe shape.

The pipe performs two functions: as the carrier of bulk material, and as a hauling means.

The typical loading section of a pipe conveyor is shown in Figure 96.

Figure 96. Rolling up of a just loaded pipe belt conveyor (BEUMER Group).

The various pipe conveyors, transported bulk material inside the rubber pipe and along the conveyor route, are shown in Figure 97, Figure 98, Figure 99.

Depending on the required strength, the special fabric belt or special steel cord belt can be used for pipe conveyors. The belt should be able to maintain the form of a circle, independent of its being empty or half-loaded or fully-loaded.

The pipe shape of the belt is unfolded to a flat belt shape before discharge, gravity take-up, and tail sections (to be re-folded further).

The length of a pipe conveyor transition section is 10 m to 15 m for a fabric belt and 20 m to 25 m for a steel cord belt.

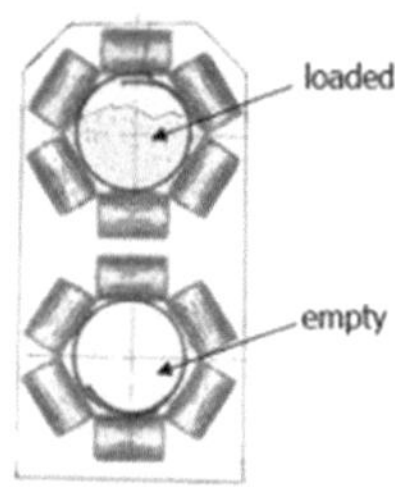

Figure 97. Cross section of pipe conveyor.

Relationship between pipe diameter, maximum recommended speed, and average spacing of supporting rollers are shown in Table 5.

The pipe conveyor rollers are usually installed as offset–type rollers: three rollers per side installed on the common roller-supporting plate (Figure 99).

Table 5

PIPE DIAMETER, mm	IDLER SPACING, m	MAX. BELT SPEED, m/s	MAX. LUMP SIZE, mm	CAPACITY. tph
150	1.2	2	50	100
200	1.2	2.2	70	160
250	1.5	2.4	85	300
300	1.6	2.5	100	500
400	1.8	2.8	150	900
500	2	3.3	200	1,600
600	2.4	3.8	250	3,000

For overland pipe conveyors, which required higher belt strength, a special steel cord belt was developed by ContiTech Group and others (Fig. 98).

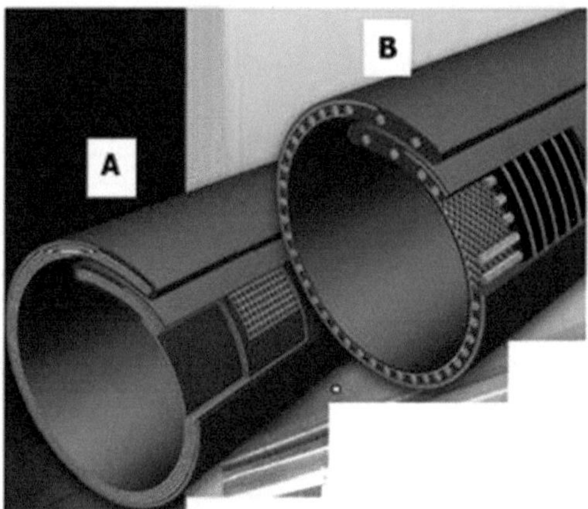

Figure 98. Fabric (A) and steel cord (B) belting for pipe conveyors developed by ContiTech Group.

Figure 99. Typical pipe conveyor (BEUMER Group).

Figure 100. Overland pipe conveyor (BEUMER Group, USA).

The advantages and disadvantages of pipe conveyors are noted below:

Advantages of pipe conveyors:

1. Bulk material is transported inside of the formed and sealed rubber tube. This is a very important consideration where environmental concerns or requirements of product purity do not allow the use of open belt conveyors.

2. There is no spillage or scattering of material from the belt return run (the return belt is also tubular formed).

3. Pipe conveyors allow smaller (than conventional trough belt conveyors) radius (*R*) in horizontal curves:

    - R = 300 × d for an EP or NN belt
    - R = 600 × d for a steel cord belt, where d is pipe diameter, m.

4. The pipe conveyor can run up a slope of +25° to +30° or downward as −20°.

5. Two-way conveyance is possible.

6. The pipe conveyor's footprint is much smaller than the footprint of enclosed gallery of a conventional conveyor.

Disadvantages of pipe conveyors:

1. The energy consumption is higher because of the higher drag resistance of the folded belt.
2. The belt is about 1.6 times wider than the belt of conventional belt conveyor at the same carrying capacity.
3. Overload (pipe conveyor should not be loaded above 70% of cross section) and oversized lumps can damage the belt.
4. The pipe conveyor may not be used to convey very hot bulk materials.
5. Spillages occur on discharge, gravity take-up and tail sections.
6. The cost of the pipe conveyor is much higher than the cost of conventional belt conveyor of the same carrying capacity.

2.1.19.4 Overland Flying Belt Conveyors

Flying Belt conveyor is a relatively new development of overland suspended conveyors. The system was patented by Agudio (a brand of Leitner, Italy, and part of Sempertrans International Group) in 2007.

The system consists of the standard trough steel cord belt supported by standard sets of garland idlers (Figure 101). The idlers are attached to and supported by anchored wire ropes. To maintain the required deflection, the ropes should be properly tensioned.

The simplified equation for calculation of a rope deflection is:

$$f = (q_r + q_m)\, L^2/8T,$$

where f is the deflection of the rope; q_r is the rope weight per m; q_m is the additional (distributed per metre) load on the rope; L is the span between two supporting piers; and T is the rope tension.

Figure 101. Typical overland Flying Belt conveyor (Agudio).

The Flying Belt conveyor differs from the trough garland idlers steel cord conveyor in the method of attaching the idlers. The sets of garland idlers of conventional belt conveyors are attached to rigid steel frames supported by line stands, installed on ground-level every 3 m to 5 m or at a height of 6 m to 10 m above ground at a span of 30 m to 40 m (Figure 102).

Figure 102. Conventional belt conveyor with garland idlers fastened to the rigid steel frame (Rulmeca, Italy).

The sets of garland idlers of Flying Belt conveyors are attached to wire ropes taut between the supporting towers.

The Flying Belt conveyor uses two pair of supporting wire ropes. The bottom-most pair of ropes support the bottom idlers, while the pair of upper ropes supports the idlers of the top belt (Figure 101). The pair of upper ropes serves also as the track ropes for the trolley (Figure 103), carrying the maintenance crew to conduct inspections, perform repairs, and carry out routine maintenance and replacement operations (Figure 104). The auxiliary ropes are tensioned under upper and under lower belts to stabilize the sets of idlers. The solid anchor blocks for initial tensioning and aligning of the ropes are installed on the head and tail stations.

Flying Belt conveyors usually run with a span of up to 1,000 m between supporting towers and at a height of up to 35 m above ground level.

Figure 104 shows the replacement of a set of garland idlers by a two-man maintenance team working from the trolley.

Figure 103. Trolley used for transporting of the inspection and maintenance team (Agudio).

Figure 104. A maintenance team replaces a set of garland idlers on a Flying Belt conveyor (Agudio).

Figure 105 and Figure 106 show an example of a Flying Belt conveyor installed in Borroso, Brazil, by Agudio, in September 2015. The transported bulk material is limestone. The length of the conveyor is 7.2 km. The belt is B1200 at speed of 4.0 m/s and the conveyor capacity of 1,500 tph.

Figure 105. Flying Belt overland conveyor (Brazil).

Figure 106. Flying Belt overland conveyor (Brazil).

Advantages of suspended Flying Belt conveyors:

1. The Flying Belt system allows for the passing over of geographical obstacles such as rivers, roads, and buildings. With spans of up to 1,000 m per section between supports, this system has a capacity that the conventional trough belt (with a span of 30 m to 50 m) and the Cable Belt system (with a span of 6 m to 8 m) do not have. In an area of difficult terrain, the suspended conveyor can significantly reduce the cost and time of civil works.
2. The Flying Belt conveyor uses the standard conveyor belt and standard sets of garland idlers.
3. The Flying Belt conveyors occupy a minimum structural footprint.
4. Low noise emission.
5. The smooth belt allows the use of standard belt cleaners/scrubbers to clean the belt after discharge and to reduce spillage on the turnover, discharge, and take-up sections.
6. The special trolley allows the maintenance/inspection team to access every point along the conveyor route.

Disadvantages of Flying Belt suspended conveyors:

1. The conveyor route should be a straight or nearly straight line. No curves or changes of conveying direction are allowed.
2. There are no walkways along the conveyor, so any inspection must be carried out by the team driving along the conveyor in a special trolley. Belt repairs, partial replacements, or broken idlers replacements, are usually carried out on the drive and tail sections, but failures along the route must be addressed by a skilled team using the trolley.
3. Flying Belt conveyors have not yet achieved the capacities and the distances achieved by trough SCB and Cable Belt overland conveyors.
4. The number of operating Flying Belt conveyors is relatively small as compared to the number of trough SCB or Cable Belt overland conveyors.

2.1.19.5 Overland RopeCon Belt Conveyors

The RopeCon belt conveyor is a relatively new technology (the first system started operations about sixteen years ago) developed by Austria-based Doppelmayr Transport Technology GmbH.

In general, the hauling tension is transferred to the belt of RopeCon conveyor by friction between the belt and the head drive pulley (or the head and tail drive pulleys).

The belting of RopeCon conveyer is a flat rubber belt, which may be a fabric-reinforced or steel cord belt depending on the required strength of the belt (conveyor capacity, and length of the transport route), with vulcanized or cold-bonded side corrugated rubber walls up to 400 mm high, and axles with two high-grade polyamide wheels (rope sheaves). The axles are bolted at regular intervals to the flat belt via fasteners (Figure 109).

The wheels run on anchored load-bearing steel track ropes, keeping the belt lifted above the ropes.

The track ropes are supported by special towers located at a distance of 1,000 m (or more) one from the other.

RopeCon uses three pairs of track wire ropes. The bottom-most pair of ropes support the bottom belt, while the pair of ropes in the middle supports the top belt. The upper-most pair of ropes give additional stability to the suspended structure and serve as the track ropes for the inspection/maintenance trolley.

The support frames are installed every 6 m to 12 m (Figure 110) to maintain the position of the track ropes along the rope spans and to retain the constant spacing between the upper and lower belts.

The frames also keep the belt in place in the case of strong side winds, preventing lift-off and de-roping.

The recommended filling of the belt is about 70% (depending on maximum lump size).

At the conveyor loading stations, the belt is supported by special slides so that the wheels are not subjected to impacts from falling bulk material.

The RopeCon can be inclined to 75° or more, but the belt should be fitted with cleats to prevent the material from running back (Figure 111).

The cleat fasteners are bolted to the flat belt and not connected to the side walls.

ContiTech AG–Continental (Germany) is one of the main suppliers of belting for RopeCon conveyors and for Flexowell conveyors.

The principle difference between these two systems (RopeCon vs Flexowell) is that the RopeCon belt is supported by axles with wheels running over track-anchored ropes, and the Flexowell horizontal carry belts are moved supported by conventional flat idlers and its return belts (without turning over) are supported by short, slightly inclined guide idlers.

Figure 107. The turning over of the return run of the RopeCon belt conveyor (Doppelmayr TT GmbH).

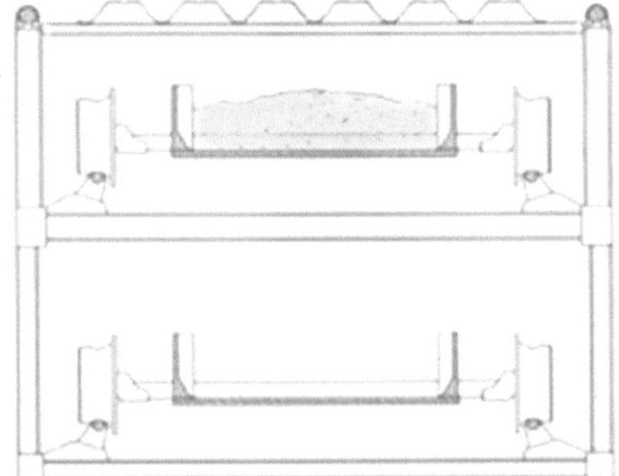

Figure 108. Cross section of RopeCon conveyor (after the return run was turned over) (Doppelmayr TT GmbH).

Figure 109. The return run of the RopeCon belt after the turning over (Doppelmayr TT GmbH).

Figure 110. RopeCon belt conveyor support frames (Doppelmayr TT GmbH).

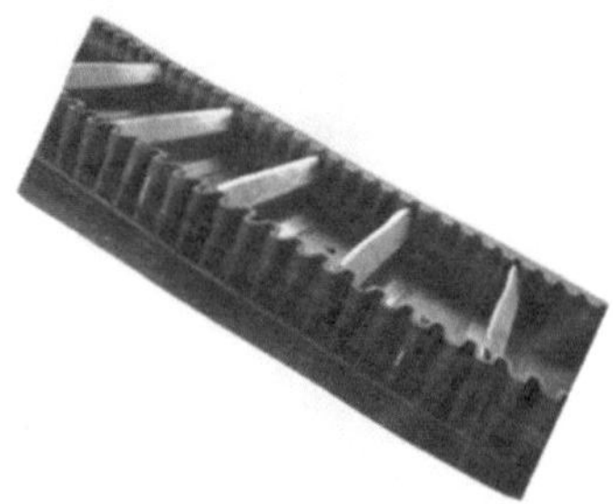

Figure 110.1. The inclined conveyor with side walls and cleats.

The typical support pier/tower is shown in Figure 111 and Figure 112. On the piers, the track wheels are lifted from the ropes and the loaded belt rolls on a number of large stationary rollers that can withstand the high belt tension and prevent overload of the track wheels at the most loaded points of the system.

Figure 111. Typical supporting piers of a RopeCon belt conveyor (Doppelmayr TT GmbH).

Figure 112. The 1.6-km-long RopeCon belt conveyor (Guatemala) (Doppelmayr TT GmbH).

Figure 113. A view of the typical RopeCon belt conveyor (Doppelmayr TT GmbH).

Figure 112 shows the RopeCon conveyor at Cementos Progreso S.A., Guatemala, 1.6 km long with a capacity 2,100 tph of limestone and a speed of 3.6 m/s, and with four intermediate towers. This RopeCon conveyor started operation in 2017.

The drive unit of the RopeCon conveyor is the typical drive unit of any conventional steel cord belt conveyor and consists of a drive pulley, an electric motor, a gearbox, couplings, and two brakes—one operational between the motor and the gearbox and another, emergency, between the gearbox and the drive pulley (Figure 114).

A typical RopeCon overland conveyor is shown in Figure 112 and Figure 113.

Figure 114. Typical drive unit of RopeCon belt conveyor (Doppelmayr TT GmbH).

As noted above, the inspection/maintenance trolley (Figure 115) runs along the special top wire ropes. The trolley can be self-moved or winch-driven.

The trolley consists of supporting frame with wheels, driving device and two maintenance platforms (one per side). The service team controls the trolley movement with "go-stop" control console, and carries out inspection, maintenance and replacement of broken parts.

Figure 115. Inspection/maintenance trolley, movable on upmost ropes.

Figure 116. Inclined covered RopeCon belt conveyor equipped with cleats.

Figure 117. An open RopeCon conveyor (Austria, Doppelmayr TT GmbH).

One of the RopeCon suspended conveyors in operation today is 3,377 m long with 1,200 tph capacity (bauxite), installed in 2007 in Jamaica. The belt was supplied by ContiTech–Continental, Germany. The drive unit of the conveyor consists of 2 x 640 kW, 480 V electric motors.

Figure 118. RopeCon belt conveyor (Mexico, Doppelmayr TT GmbH).

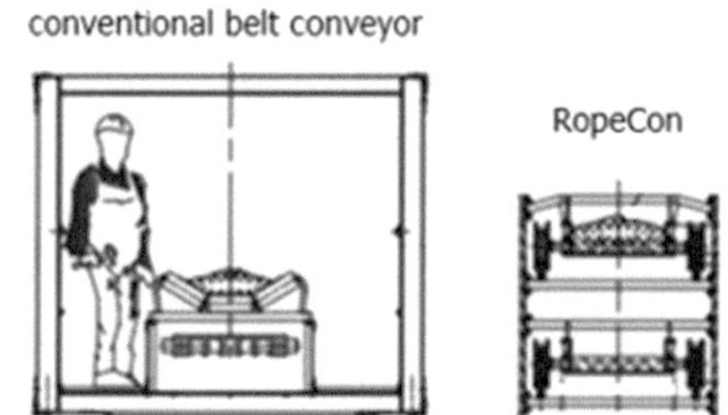

Figure 119. Comparison cross sections of conventional and RopeCon conveyors.

Advantages of RopeCon conveyors:

1. The RopeCon conveyor allows the passing over of geographical obstacles such as rivers, roads, and buildings, with spans of 1,000 m and more between neighbouring supports, whereas the trough SCB and the Cable Belt overland conveyors do not have this capability. In areas with difficult terrain, the RopeCon conveyor can significantly reduce the cost and the time of civil works (Figure 112, Figure 118).

2. The RopeCon conveyor occupies a minimum structural footprint. The comparison between cross sections of a conventional belt conveyor and the RopeCon conveyor is shown in Figure 119. RopeCon conveyors have been built without walkways, using a trolley for maintenance and parts replacement.

3. Low noise emission.

4. According to the manufacturer, because of a low running/rolling resistance, the friction energy consumption of the RopeCon conveyor is by 30% less than the energy consumption of a conventional belt conveyor of the same capacity and following the same route.

5. Turning over the belt (Figure 108) eliminates spillage along the conveyor route.

6. The loading and discharge stations of the conveyor are fixed points for monitoring, checking, and replacing of wheels and axles passing through.

Disadvantages of RopeCon conveyors:

1. The construction of the belt doesn't allow the use of standard belt cleaners. Even if the belt was turned over, the discharge, the take-up, and the turnover sections (especially when the belt is fitted with cleats and/or when it transports wet or sticky bulk material) it causes spillage. The spillage takeaway belt or drag chain conveyor arrangements should be designed and installed at the very beginning of the RopeCon's conveyor operation. Efficient cleaning of spillage is an area that needs further development and could involve such solutions as belt rapping and washing.

2. The conveyor route must be a straight or nearly straight line. No changes of direction are allowed.

3. There are no walkways at sides of the conveyor, so inspection and maintenance must be carried out by a two-man team driving along a conveyor in the special trolley. Belt repairs, partial replacements, or broken wheel replacements are usually carried out on the drive and tail stations, but in case of failure that must be treated on the spot, these serious technical problems require the skilled team coming in and operating from the trolley.

4. The RopeCon system has not yet achieved the distances of the conventional SCB overland conveyors or Cable Belt conveyors.

5. High Capex.

2.1.19.6 Overland Ropeway Conveyors

Doppelmayr (Austria) ropeway conveyors for bulk materials are discontinued tailor-made conveying systems which, in principle, were based on the Doppelmayr reversible aerial tramways that can be found at any ski resort. Instead of passenger cabins, the open hoppers, carrying bulk material, are fastened to and moved by steel ropes (Figure 123, Figure 124).

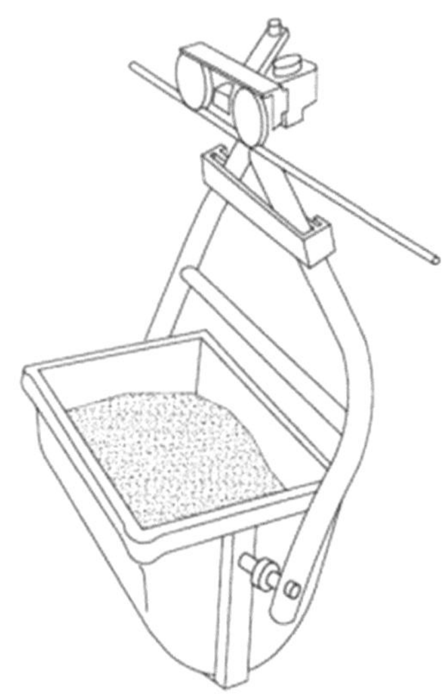

Figure 123. Ropeway conveyor: open hoppers carrying bulk material.

Figure 124. Example of an operating ropeway conveyor.

Advantages of overland ropeway conveyors:

1. The conveyors are used to suit the specific local requirements where other overland transporting systems (Flying Belt conveyors, RopeCon conveyors, and so on) are unable technically or economically resolve the transport problems (e.g. hill crossing or passing inaccessible areas).
2. The maximum capacity of such conveyors is up to 500 tph with a rope span of 1.5 km or more.
3. The maximum length of a ropeway conveyor is 13.5 km (Damodar, Maharashtra, India).

Disadvantages of overland ropeway conveyors:

1. Interrupted feeding/unloading.
2. Capacity limitation.
3. Dependence on weather (e.g. during strong side winds or heavy rains the conveyor will be stopped).
4. Hoppers must be cleaned after transport of wet or sticky bulk materials.
5. Dust generation during discharge of dusty bulk materials.
6. High Opex and high Capex.

2.2 Bucket Elevators

Bucket elevators are a simple and reliable means for lifting of bulk materials of the required capacity to the required height [22, 23]. The elevators are also used for S-conveying of bulk materials: they can be fed by a belt feeder/feed conveyor, lift the material and transfer material onto a takeaway belt conveyor (Figure 125) and also can be manufactured in Type S or Type C variations.

A bucket elevator consists of a drive unit, pulleys for belts or sheaves for chains, housing, and a serious of buckets attached to the endless belt or chain(s).

Calculations of friction of a bucket elevator include such additional component as cup digging.

Parameters of a bulk material play a significant role in the reliable operation of a bucket elevator. Only free-flowing bulk materials (no wet or sticky, and no sludge!), are recommended for lifting using bucket elevators. It is extremely important to feed the elevator at a continuous and consistent rate and to evenly distribute the material across width of the bucket. The recommended rate of filling of a bucket is 65%÷75%.

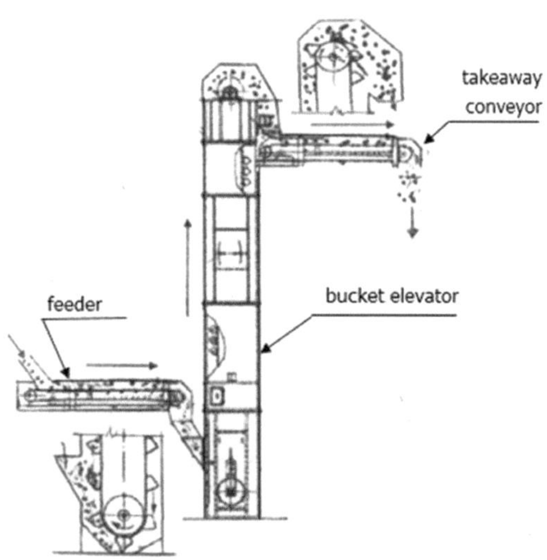

Figure 125. Principle scheme of operation of a bucket elevator.

Bucket elevators are divided into centrifugally discharged and continuous (gravity-discharged).

Centrifugally discharged *rubber belt* elevators (Figure 126, A) are elevators, operated at a high belt speed of 2.0 m/s to 3.5 m/s, at a capacity up to 600 tph, and with a lift of 50 m and more. The buckets are attached (bolted) to the belt (Figure 128, Figure 129).

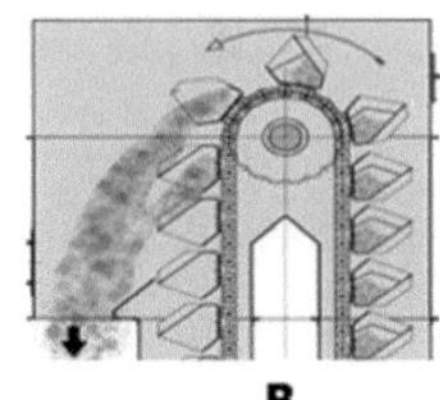

Figure 126. Centrifugal (A) and continuous-gravity (B) discharge of bucket elevators (4B Group, UK).

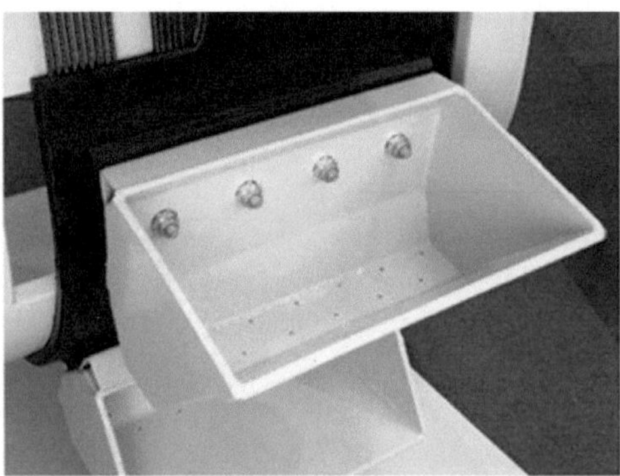

Figure 127. Buckets connected to the steel cord belt with bolts (BEUMER Group).

Continuous or gravity-discharged *chain* elevators (Figure 126, B; Figure 128; Figure 129) operate at slow speed of 0.5 m/s to 1.2 m/s (the speed of the chain is limited by the dynamic fluctuations of the chains at higher speeds), using two parallel chains with buckets attached to chains very close to each other. The capacity of a standard elevators can reach 600 tph and more, and the height of 60 m and more. Gravity-discharged elevators can transport lumps up to 120 mm.

Figure 128. Drive sprocket of roller chain bucket elevator (A) and drive pulley of belt bucket elevator (B) (BEUMER Group).

All bucket elevators operate within enclosed housings and this causes difficulties in detecting mechanical breakdowns, so monitoring of the elevator during its operation is highly recommended. The 4B Group, for example, proposed special devices for monitoring belt slippage, belt misalignment, pulley misalignment, and bearing failure. Figure 130 shows an example of the monitoring of a bucket belt elevator.

Figure 129. Attachment of a bucket to the round link chain (4B Group, UK).

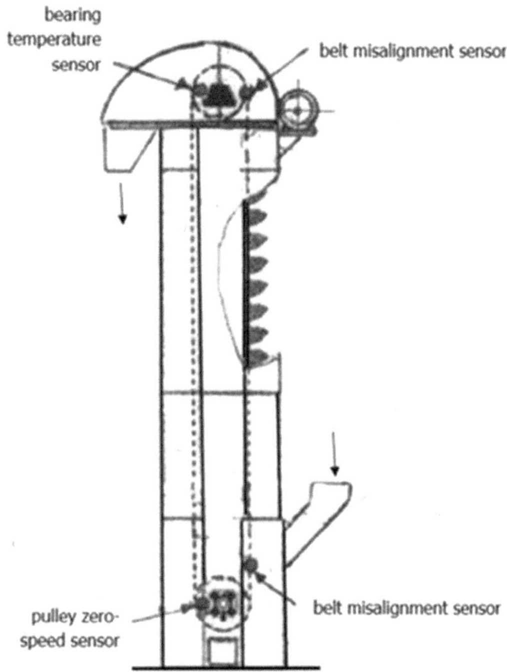

Figure 130. Recommended monitoring of a belt bucket elevator (4B Group, UK).

Figure 130.1. Chain elevator lifting 600 tph of lime (Bedeschi MW Conveyors).

As noted above, the choice of the optimal type of the elevator depends on characteristics of bulk material, capacity, required height and on the environment.

CONSIDIRATIONS REGARDING BULK MATERIAL CHARACTERISTICS:

CENTRIFUGAL ELEVATORS.

These belt elevators are mainly used for following bulk materials:

- Coarse and granular materials
- Corrosive bulk materials
- Materials with relatively high moisture content
- Materials without big lumps, for handling bulk materials with a lump size up to 25 mm (< 10%).

CONTINIOUS ELEVATORS.

These elevators are mainly used for following bulk materials:

- Hot bulk materials
- Bulk material containing big lumps (up to 120 mm)
- Fragile materialsDusty materials

Advantages of bucket elevators:

1. Enclosed upward (90°) lifting of bulk materials.
2. Small footprint.
3. Chain elevators can be used for hot bulk materials.
4. Large capacities and significant heights are available.
5. Simplicity and reliability.
6. Low-maintenance requirements.

Disadvantages of bucket elevators:

1. The rubber belt is usually lasts only three to four years. The belt is damaged during cup digging of spilled bulk material at the bottom of the housing and by the wrapping around pulleys, especially in places where buckets are bolted to the belt (Figure 127).
2. Belt slippage can cause clogging of the housing.
3. Elevator, transferred dusty materials, should be equipped with a dedusting system.
4. Monitoring of the elevators is highly recommended (Figure 130).
5. High Capex.

2.3 Screw Conveyors/Feeders

Screw conveyors are widely used for the enclosed transportation of bulk materials on short (4 m to 8 m), horizontal, inclined and vertical routes [24, 25]. Longer horizontal distances require installation of intermediate bearings or a number of individual conveyors should be connected inline feeding one another.

Screw conveyors can be divided into two main groups:

- screw conveyors
- screw feeders.

Screw conveyors usually transport bulk materials at a filling rate of 20%÷45% and on relatively short distances (from 4 m to 8 m) to avoid the use of internal bearings operating amid bulk material and requiring regular lubrication and maintenance. The intermediate internal bearings are installed in the housing (Figure 135) every 3.0 m ÷4.0 m.

Figure 131 shows example of a screw conveyor design: capacity Q≈20 tph, length L ≈ 8 m, speed v = 65 rpm, D/d = 270 mm/168.3 mm, angle of spiral inclination $\alpha = 55^{\circ}$, filling rate $\varphi = 35\%$, without intermediate bearings.

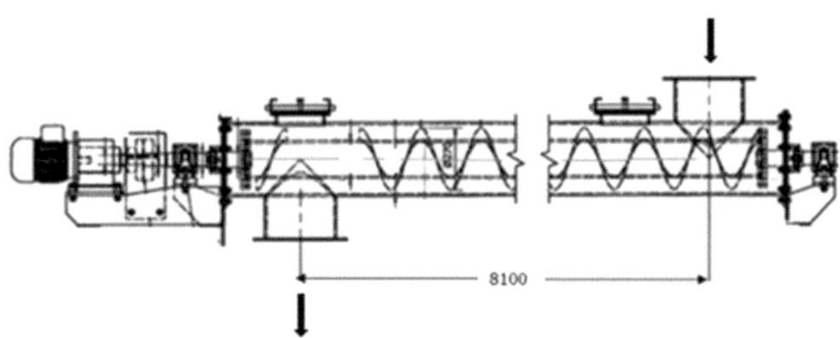

Figure 131. Example of screw conveyor design.

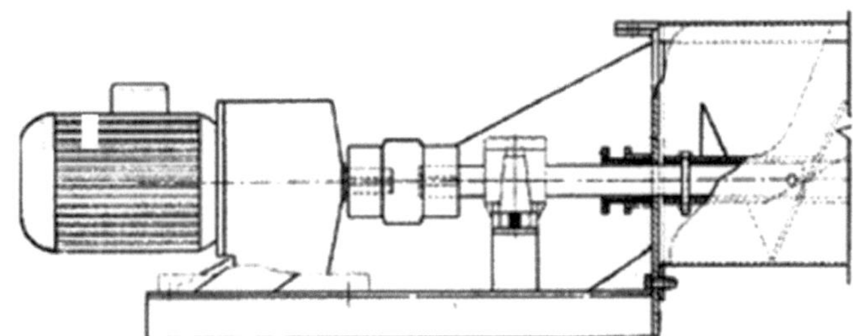

Figure 132. Typical assembly: connection of the drive unit to the screw.

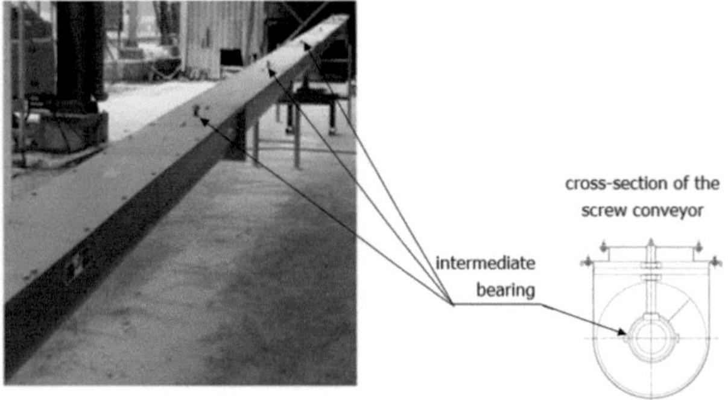

Figure 133. Long slope screw conveyor with intermediate bearings.

For a standard screw conveyor 3.5 m (≈10 feet) long, the recommended shaft diameter is 114.3 mm (4″). For longer screw conveyors, it is important to eliminate the excessive shaft deflection and to keep the recommended [24, 25] minimum spiral/housing gap of about 1/2″, so the shaft diameter should be increased. It's worth paying attention that shown above, 8m long screw conveyor (Figure 131) was designed without intermediate bearing, but instead of this, the diameter of the shaft was increased to 168.3 mm.

Figure 134. The vertical screw conveyor of the Siwertell ship unloader.

The principle (the interaction of material with the spiral and with the housing) of the screw conveyor slightly differs from the principle of screw feeders.

Screw Conveyors

For normal operation of the screw conveyor, *it is necessary the friction between the spiral and the material should be less than the friction between the material and the housing; otherwise, the material will be rotated along with the spiral instead of slipping along the spiral surface to be pushed forward by the rotating spiral.*

The normal filling rate of a standard screw conveyor is relatively small, 20% to 45%. This low filling rate of material can be explained by the need to keep a reduced weight of material on the spiral and, thus, reduce friction between the material and the spiral. The choice between a 20%, or 30%, or 45% filling rate depends on the material–steel friction coefficient and the bulk density of the bulk material. *The more "material–steel" friction coefficient and the more bulk density (and, hence, weight) of the material, the less the recommended filling rate of the screw conveyor.*

Vertical screw conveyors (Figure 134) operate at a speed of about 500 rpm (instead of the generally accepted for horizontal screw conveyors speeds of 40 rpm100 rpm) and it is necessary to create significant centrifugal force between the material and the housing. The centrifugal force increases friction between material and housing and prevents rotating of the bulk material with the spiral. The rotation of the vertical spiral *relative to non-rotating material* moves the mass of material upward to the discharge chute.

Screw Feeders

Horizontal screw feeders are usually installed under the discharge outlets of hoppers and silos to convey the bulk material across short distances (up to 5 m or 6 m) at initial filling rate of about 100%.

The 100% filling increases the friction between the bulk material and the spiral. So, the *additional vertical pressure on the feeder is required* to increase friction between the material and the housing and to prevent the bulk material from rotating along with the spiral. The weight of the material in the hopper *above the feeder, ensures this pressure* and produces the required bulk material/housing friction allowing *non-rotating* material to be moved forward by the rotating spiral.

By the way, now we can realize why, using a meat grinder, we push the minced food downward – to increase pressure and, thus, friction between the food and the grinder housing, and thus *prevent rotation of food along with a horizontal screw of the grinder!*

In order to continue the movement of the material already *outside the hopper outlet and thus without* material pressure from the hopper, the feeder must begin to operate as a conventional screw conveyor. For this it is necessary to increase the diameter, pitch and angle of inclination of the spiral (Figure 135). These changes allow reducing the filling rate from the initial 100% to accepted for conventional screw conveyors filling of 20%÷45%.

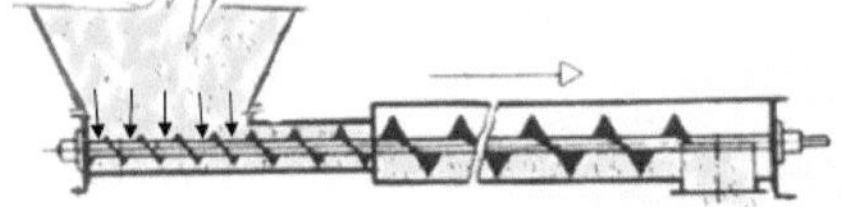

Figure 135 Scheme of screw feeder/screw conveyor.

For hoppers with long slot openings, the problem is how to discharge the bulk material evenly along the slot without "dead" zones.

One of the practical solutions of the problem is the installation of a cone-type screw feeder (Figure 136). The discharge rate of such a feeder is increased along the hopper opening: the smaller the ratio *d/D* (*d* and *D* are the shaft and spiral diameters respectively), the higher the feeder capacity. So, the feeder increases its capacity along the slot and, as a result, the hopper is discharged evenly.

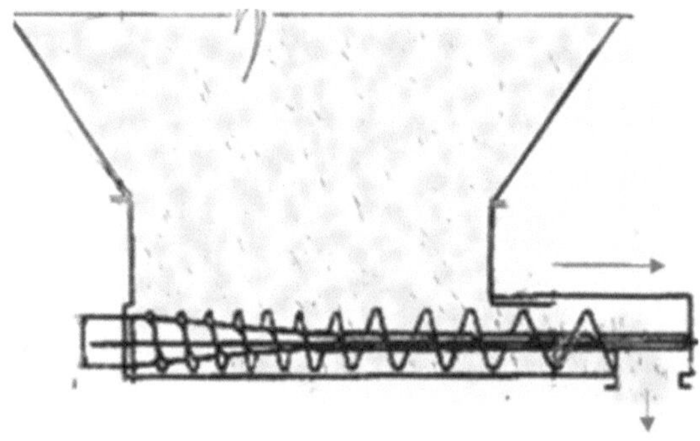

Figure 136. Scheme of cone-type screw feeder.

The main parameters required to choose the right screw conveyor:

1. Characteristics of the bulk material (bulk density, flowability, friction coefficients), capacity, feeding, and total conveying distance.

2. φ - filling rate of the screw as a function of friction between the bulk material and the housing, and the friction between the material and the rotating spiral:
   - Non-free-flowing material with material–steel friction coefficient 0.4÷0.6, φ = 0.15÷0.20.
   - Free-flowing material with material–steel friction coefficient 0.30÷0.4, φ = 0.25÷0.30.
   - Free-flowing material with material–steel friction coefficient 0.2÷0.3, φ = 0.3÷0.35.
   - Free-flowing material with material–steel friction coefficient 0.15÷0.2, φ = 0.33÷0.45.

   As noted above, the higher friction coefficient between material and spiral, the less filling rate in order to reduce friction between the material and the spiral and to prevent the material from rotating along with the spiral.

3. Capacity reduction factor *K* depends on the angle of conveyor inclination, β:

   $\beta = 0^\circ$ K = 1.00

   $\beta = 5^\circ \div 10^\circ$ K = 0.85

   $\beta = 10^\circ \div 15^\circ$ K = 0.75

   $\beta = 20^\circ$ K = 0.55÷0.65

The same principle that was noted: the greater the slope angle of the conveyor the lower filling rate is recommended (the reduced weight of material *on* a spiral lessens material/spiral friction, in comparison with material/housing friction, and thus prevents the material from rotating together with the spiral).

4. Rotation speed (r) as a function of the screw diameter (D) is shown below:
   - Screw diameter D = 160 mm ÷ 200 mm, speed r = 25 rpm÷120 rpm.
   - Screw diameter D = 250 mm ÷300 mm, speed r = 20 rpm÷80 rpm.
   - Screw diameter D = 400 mm – 500 mm, speed r = 20 rpm÷60 rpm.

5. Reduction of pitch of screw accordingly decreases the vertical component of the material load on the spiral and thus reduces friction between material and spiral.

Conclusion:

For normal screw conveyor and screw feeder operation, it is necessary to maintain a more friction forces between the material and the housing than between the material and the spiral. Otherwise, material will start to rotate alone with the spiral and this will lead to blockage of the screw conveyor.

A typical long and slightly inclined screw conveyor with intermediate bearings is shown in Figure 137.

Figure 138 shows a typical screw conveyor hanger bearing, which can be of several types: ball or roller, bronze-bushed, Babbitt, oil-impregnated wood, and so on. These bearings operate inside the moving layer of bulk material that significantly reduces their reliability and service life. The bearings require regular inspection and manual/automatic re-greasing.

Below are the CEMA [23] recommendations for inclined screw conveyors:

> "Several things can be done to overcome many of the problems associated with an inclined screw conveyor:
>
> - Limit the use of standard components to inclines of less than 25°, preferably not over 15°.
> - Use the close clearance between trough and screw.
> - Increase the speed over that applicable for a horizontal screw conveyor of the same size.
> - Use short pitch screws, 2/3 or 1/2 pitch".

The various types of screw spirals are shown below:

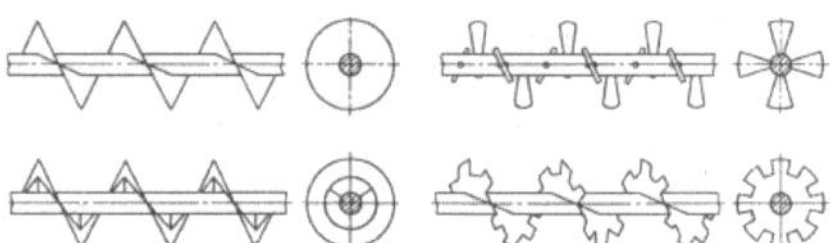

Example from practice:

A screw conveyor automatically distributing the dust from one baghouse filter between three big bags in turn is shown in Figure 137. The diaphragm pressure sensors, installed on outlet pipes of screw conveyor detect when the bulk material has filled up the big bag and sends a corresponding signal to the control room of the terminal to replace the filled bags with the empty ones.

Figure 137. Screw conveyor automatically distributing dust from the baghouse filter between three big bags.

Advantages of screw conveyors:

1. The main advantage of screw conveyor is its ability to convey bulk material inside a sealed pipe (or sealed housing) without spillage and dust emission.
2. Special screw conveyors can be used for vertical transportation at a speed of about 500 rpm and capacities up to 1,000 tph (Siwertell).
3. Simple, reliable, and low-maintenance equipment.

Disadvantages of screw conveyors:

1. Upwardly inclined screw conveyors significantly reduce their capacity. For example, the slope by 20° reduces the nominal capacity by 30%÷50%, depending on the characteristics of the bulk material.
2. Relative low capacities of conventional screw conveyors because they are limited by speeds of 70 rpm to 80 rpm for small diameters and 40 rpm to 60 rpm for larger diameters. The filling rate of conventional screw conveyors are relatively small and amounts to 20%÷45% (depending on the characteristics of the bulk material).
3. Long screw conveyors (L ≥ 8 m) require installation of intermediate suspended bearings operating inside the moving bulk material. Such bearings require regular inspection and frequent replacement.

2.4 Drag Chain Conveyors

The en-masse drag chain conveyor (Figure 138) drags bulk material on the bottom of the enclosed trough with "flights" (flat horizontal plates) fixed to two sides of an endless single strand chain [25, 26].

Figure 138. Typical en-masse drag chain conveyor (4B Group, UK).

The principle of the drag chain conveyor operation:

The bulk material is loaded via feed chute from above, passes through the upper run of chain with flights, and falls on the bottom of the trough, where flights of the lower run of the chain, drag the material to the discharge opening(s) at speed of 0.1 m/s to 1.0 m/s.

The attachment of the flights to the link (A) and to the roller (B) drag chains are shown below (Figure 139).

Figure 139. Attachment of flights to link chain (A) and to roller drag chains (B) (4B Group, UK).

The tail section of the drag chain conveyor is equipped with a screw chain-tensioning system.

The en-masse drag conveyor can operate at high temperatures (+400°C) and at capacities of up to 1,000 tph.

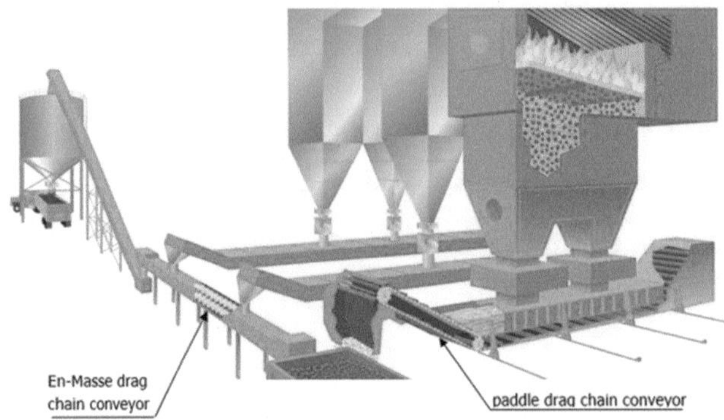

Figure 140. Using drag chain conveyors for hot materials (Metso).

Figure 141. One-chain en-masse drag conveyor (4B Group, UK).

En-masse drag conveyor operation: the skeleton flight configuration and short flight pitch cause the bulk material to move in a solid horizontal column as flights do not contact the bottom of the trough. If the height of the material column is much higher than the calculated height (in principle, bulk material can occupy as much as 90% of the trough's cross section), when self-regulation of the capacity takes place: the *shear stress*, caused by friction of the material with the side walls of trough becomes greater than the *strength* of the bulk material in the column (depending on the internal frictional angle of the material). As a result, the upper layers of the material column are braked by the friction and stop. The following slippage between layers results in intensive wear and in the loss of conveyor capacity. So, the maximum capacity of the an en-masse drag conveyor is limited by the parameters of bulk material.

The single-chain dragging allows conveying along a slightly curved horizontal route. Different configurations of drag conveyor skeleton flights are intended for various bulk materials and for various angles of inclination of the conveyor and are usually recommended by a drag chain conveyor manufacturer.

Paddle drag conveyors usually operate with two endless chains and with paddles (horizontal flat plates) attached between the two chains (Figure 139, A). The height of the material layer is slightly higher than the height of flats (the material can occupy about 50% from the cross section of the trough). So, in this case, the low horizontal column of material is dragged by flights having about the same cross section as the cross section of the dragged bulk material.

The paddle-drag conveyors are more power-consuming and their capacity is lower than the capacity of en-masse conveyors.

For a long conveying distance or to change the direction of transporting, a number of individual drag conveyors can be installed in sequence and connected with fully enclosed transfer points.

As drag chain conveyors are enclosed transporting systems, so monitoring of the conveyor operation is highly recommended. An example of the monitoring is shown in Figure 142.

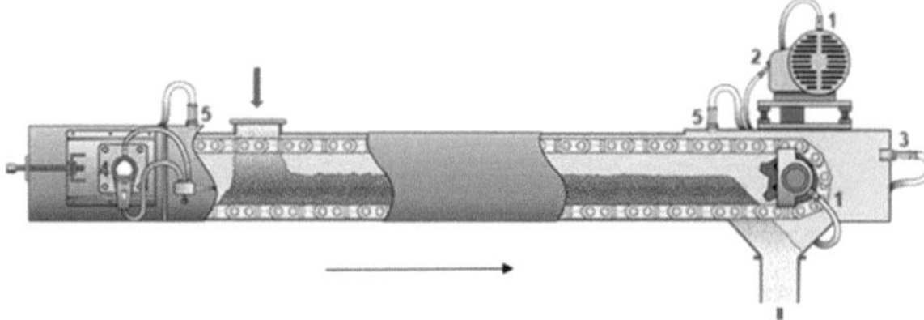

Figure 142. Monitoring of drag chain conveyor, where 1 is the bearing temperature sensor,3 is the surface temperature sensor, 3 is the plug/choke sensor, 4 is the speed sensor, and 5 is the slack/ broken chain detector (4B Group, UK).

Advantages of en-masse drag chain conveyors:

1. Totally enclosed transportation, no spillage, no dust emission.
2. No problem with conveying high-temperature products.
3. Long conveyor routes—up to 100 m.
4. Wide range of capacities, from 1 tph to 1,000 tph.

Disadvantages of en-masse drag chain conveyors:

1. Attrition of granular materials begins at speed higher than 0.5 m/sec.
2. Accidentally hit a piece of metal or foreign metal object can destroy this fully enclosed chain conveyor.
3. High power consumption: most of power consumption spent on friction occurs during dragging of chain(s) and flights along the trough.
4. The Capex is relatively high.

Special drag chain conveyors differing from those described above, are often used in mines. They are manufactured by Mackina Westfalia (Spain) and others. The conveyors transport lumpy bulk materials (ore) on their *top* run and can be installed under 300 ÷ 400 tonne pile of ore (Figure 143, Figure 143.1).

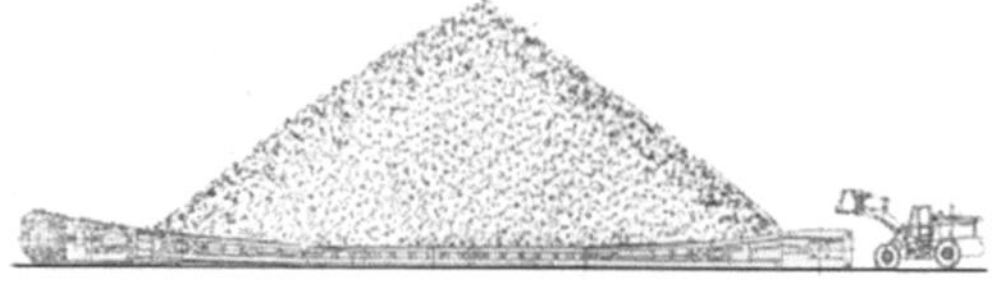

Figure 143. Drag chain conveyor, operating under a pile of ore in a mine.

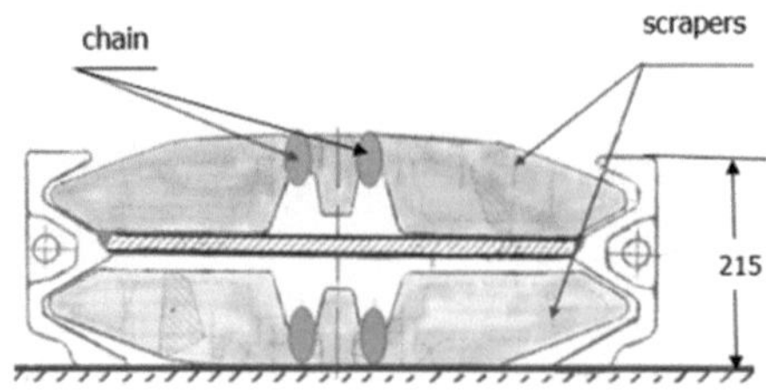

Figure 143.1. Cross section of the mine drag chain conveyor.

2.5 Flexowell Conveyors

Flexowell elevating systems are used mostly for S-shape conveying routes. The Flexowell conveyor consists of a rubber belt with rubber corrugated side walls, rubber cleats, deflection wheels, supporting idlers (on the horizontal sections) and a drive unit (Figure 144, Figure145).

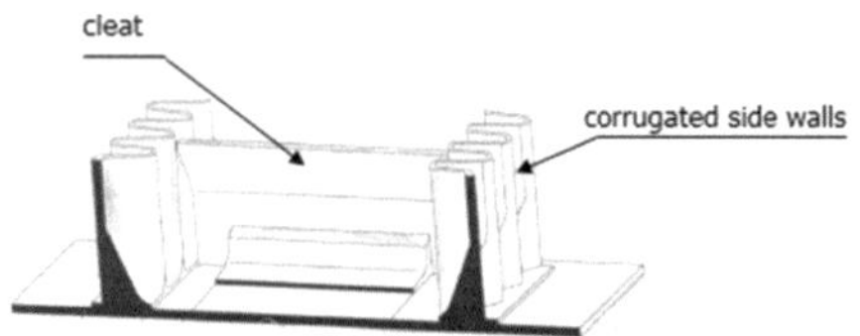

Figure 144. Cross section of the upper run of a Flexowell conveyor (ContiTech–Continental).

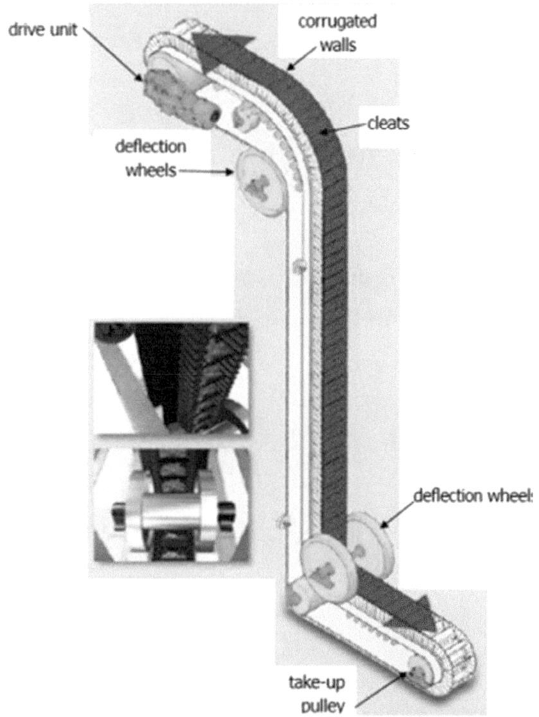

Figure 145. Principle scheme of Flexowell S-type conveyor (ContiTech–Continental).

The driving force moving the Flexowell conveyor is transferred to the belt by friction between the drive pulley and the belt (the same principle that drives conventional belt conveyors).

Hence, Euler's law and resistance-summarizing equation can be used for calculations of Flexowell S-type elevators (see section 2.1.4).

Various configurations of Flexowell conveyors are shown below (Figure 146):

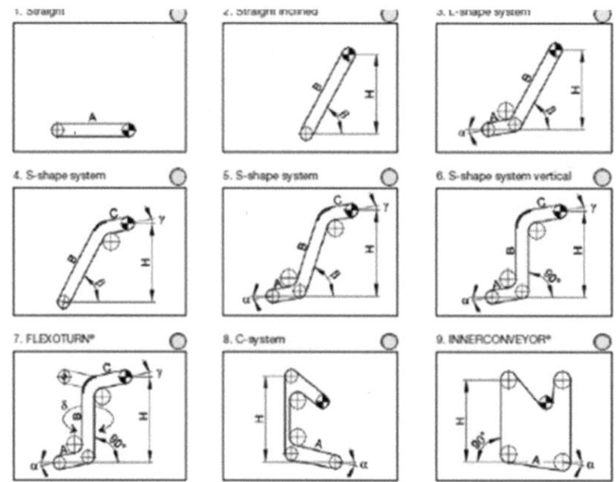

Figure 146. Various configurations of Flexowell conveyors (ContiTech–Continental).

Advantages of the Flexowell conveyor:

1. 1. Possibility to elevate a bulk material to a significant height (up to 90 m) at high capacity and without material running back problem.
2. 2. The lower horizontal feeding section, the elevating section, and the upper horizontal/slightly inclined takeaway section are arranged in one uninterrupted transporting S-type conveying system with one common drive unit.
3. 3. Corrosive and abrasive bulk materials can be transported without problem.
4. 4. Reliable, low-maintenance conveying.

Disadvantages:

1. The main disadvantage is the spillage of material from return belt. Loose material stuck between side walls and the cleats and pours out when the belt returns for loading. The belt cannot be cleaned off by any standard belt cleaner/scraper. The rapping cleaners help to clean belting but they are not good enough. So, careful consideration should be given to solve the spillage problem at the preliminary stages of a project.
2. The use of inclined Flexowell conveyors is very uncommon. The maintenance team will find it difficult to climb up an inclination higher than 15°.
3. High Capex.

2.6 Tubular Drag Conveyors

Tubular drag conveyors (Figure 147) are used for transporting light and friable bulk materials (cereals, almonds, beans, coffee, rice and so on) through enclosed tube where an endless single steel chain (or rope) drags the discs attached to the chain at regular intervals. The material is dragged between and by the discs at low speed and horizontally, inclined and vertically in one system.

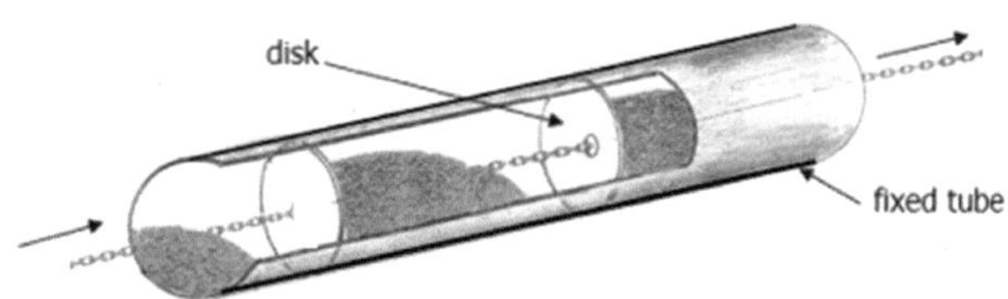

Figure 147. Work principle of tubular drag conveyor.

Advantages of tubular drag conveyors:

1. The ability to transport very small amounts of light granulated bulk materials.
2. Minimize breakage and degradation of friable products.
3. Combine bidirectional transport with multiple inlets and outlets.

Disadvantages:

1. Limited list of friable granulated materials recommended for conveying. Dusty, wet, polydispersive bulk materials clog the tube.
2. Low capacity of conveying (maximum diameter of transport tube is 152.4 mm at a conveying capacity up to 25 tph).
3. Low transporting distance (up to 50 m).

2.7 Sandwich-type Belt Conveyors

A sandwich-type belt conveyor is used for S-type transporting of bulk material using standard belt conveyors. The goal is achieved by using a second belt conveyor (installed above the material carrying conveyor) whose return belt moves in the same direction and with the same speed as the lower conveyor, and compresses the conveyed material between two belts. The pressure prevents the material from running back (Figure 148).

The pressure can be achieved as by mechanical means (Figure 147, A) or by compressed air (Figure 147, B).

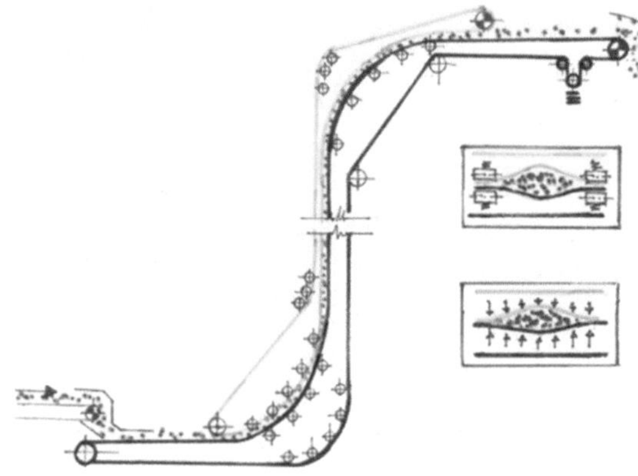

Figure 148. Principle scheme of sandwich-type belt conveyors.

Advantage of the sandwich-type conveyor:

The only advantage of this conveyor is the use of standard belt conveyors for S-type transportation.

Disadvantages:

1. 1Complicated route of belt conveyors different from usual conveyor routes and requires a complex, expensive support structure to ensure the approach of maintenance personnel for inspection, cleaning and replacement of broken parts.
2. Replacing the belt is a difficult and time-consuming operation.
3. The transported bulk material should be homogeneous, without big lumps.
4. The feeding should be continuous and consistent.
5. High Capex.
6. Extremely high Opex (compared to Flexowell and other similar S-type conveyors).

2.8. Feeders

Feeders are the short conveying devices that are usually installed under outlets of hoppers, silos, and bins, and serve for the transfer the stored bulk material with a certain capacity onto the takeaway belt conveyor, crushers, screeners, and so on. The most widely used feeders are belt feeders, apron feeders, vibratory feeders, and screw feeders (see section 2.3).

2.8.1 Belt Feeders

Belt feeders are short belt conveyors that are used to discharge hoppers, bunkers so on.

The principle differences between a belt feeder and a conventional belt conveyor are shown below:

- Belt feeder

The low-speed (0.5 m/s ÷ 1.0- m/s).

Belt feeder carries the considerable bulk material headload (which depends on the cross section of the hopper outlet, the internal friction angle of the bulk material, the height of the hopper, and so on) both during conveying and conveying interruptions. To start its movement, the feeder must overcome the friction forces between the layer of material resting on the feeder and the material located above this layer within the hopper.

- Belt conveyor

The high-speed (1.5 m/s ÷3.0 m/s).

Belt conveyor should be fed with a pre-calculated capacity (ton per hour). The feeding chute should be smoothly curved in direction of belt movement, to eliminate impact of falling bulk material and to promote the acceleration of the material.

Note:

Use the belt conveyor as a feeder leads to rapid wear of the belt, spillage, dust emission, so on, and is completely not justified from the technical and operational points of view.

The belt feeder capacity is determined by the cross-section of the layer of bulk material on the outlet of the hopper and the speed of the feeder (Fig. 149).

To overcome significant friction force, the belt feeders are built as low-speed powerful devices.

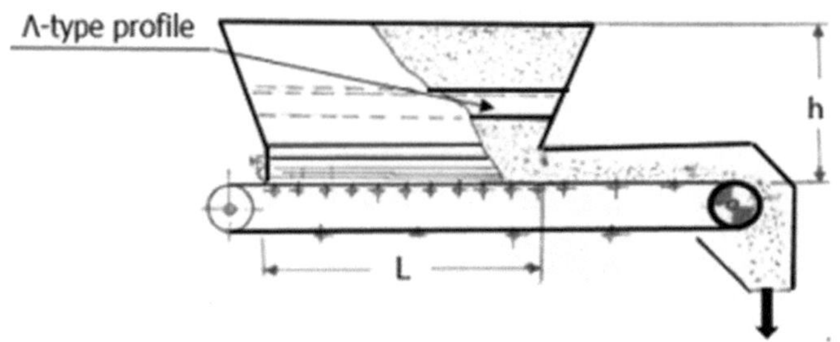

Fig. 149 Principle scheme of belt feeder.

One of the equations used to calculate pull-out force (*F*) of the belt feeder is shown below (Bruff's method):

$$F = 2\,L^2 \times b^2 \times f \times \varrho \times h \times n_{st} / [1{,}000 \times (c + b)],$$

where *F* is the required pulling force; *L* is the length of hopper's bottom opening; *b* is the width of the hopper's bottom opening; *h* is the effective material height ($h = 2b$); ϱ is the bulk material bulk; *f* is the material–internal friction coefficient; and n_{st} is the starting experimental coefficient, $n_{st} = 4$.

The several practical recommendations for the design of a belt feeder, based on many years of the experiences, are shown below:

1. To reduce the dynamic impact and the static pressure of the loaded bulk material on the feeder, the receiving hopper that loaded by shovels, tipper trucks, or bottom-dumping railway wagons, should be equipped with internal deflection plates or Λ-type profiles (Figure 150, Figure 151).
2. The carry belt of the feeder is to be supported by sets of troughing 35° rubber-coated idlers installed (under the hopper outlet) very close to one another with a gap of 30 mm to 50 mm.
3. The skirt boards internal width $b = 0.5$ B, with a height of up to $h = 0.5$ m.
4. Low speed: the recommended speed of belt feeders is between 0.5 m/s and 1.0 m/s.
5. The required power for most of belt feeders (after completion of recommendations no. 1 to no. 4) is between 15 kW and 37 kW (depending on the parameters of bulk material).

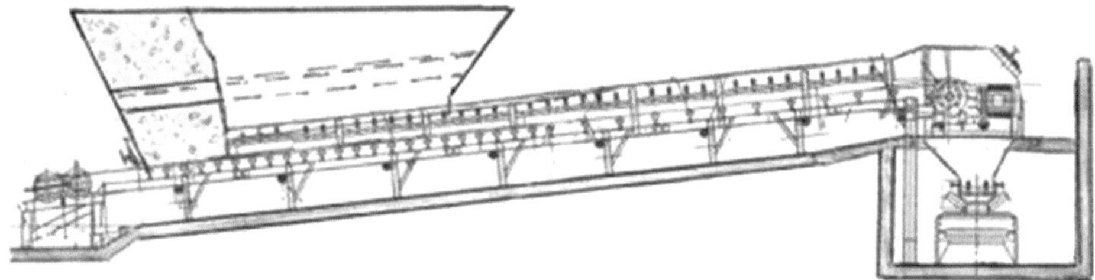

Figure 150. Example of belt feeder.

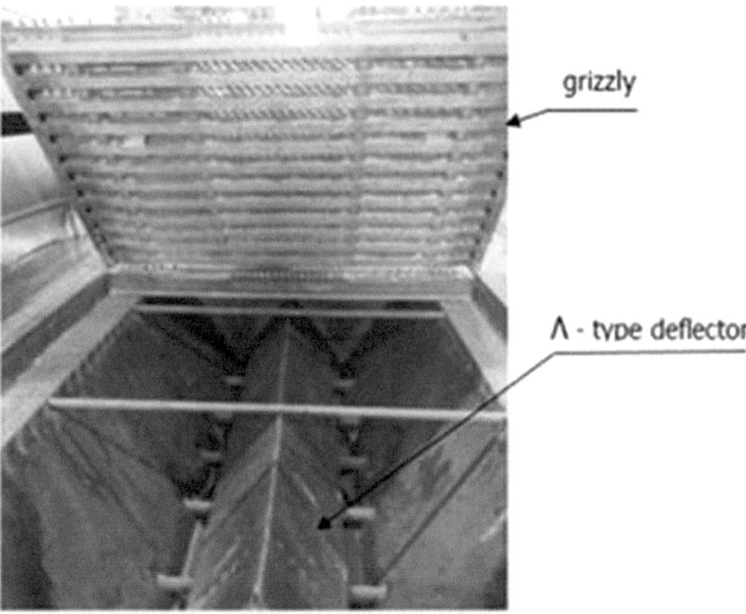

Figure 151. The discharge pit with a grizzly raised to show the Λ-type profile that was installed inside the discharge hopper to reduce material the impact and static pressure on belt feeder.

If the distance between back wall of the hopper and the tail pulley of the feeder is not enough for belt transition distance, a belt with corrugated walls is the acceptable solution.

The receiving hoppers with belt feeders are shown in Figure 152 and Figure 153.

Figure 152. The 15 m³ discharge hopper with belt feeder transferring bulk urea into a chain elevator loading a takeaway belt conveyor (Finland).

Figure 153. Belt feeder installed under movable grab-loaded hopper (Finland).

2.8.2 Apron Feeders

Apron feeders are usually used when lumpy, sharp edged, abrasive or sticky bulk materials are discharged by large tipper trucks. In order to operate in such hard conditions, the apron feeder, installed under the receiving hopper, should be robust and strong enough to withstand the impacts (Figure 154, Figure 155, Figure 156) and to transfer the material into the crusher, lump breaker, or similar equipment with further loading of takeaway belt conveyors.

The feeder consists of two heavy-duty chains with fabricated pivoted steel pans bolted directly to the chains (Figure 154, Figure 155) and a low-speed, high-torque drive (controlled by the variable frequency drive).

The pans are overlapped to provide leak resistance on the carry side of the feeder.

The start-up pulling force of the feeder can be twice as high as the running force.

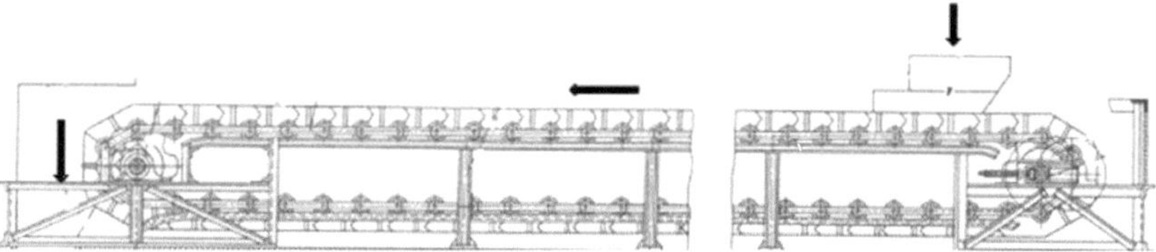

Figure 154. Example of an apron feeder design.

Figure 155. Typical apron feeder.

Figure 156. Drive sprocket of chain apron feeder (BEUMER Group).

Figure 157. Apron feeder installed under receiving hopper.

Advantages of apron feeders:

1. Built to withstand impacts of falling heavy, lumpy bulk materials.
2. Can transfer sticky, non-free-flowing bulk materials.
3. Can convey hot bulk materials (up to +400°C).
4. High capacities of 1,000 tph and more.
5. Reliable, low-maintenance equipment.

Disadvantages:

1. High energy consumption.
2. Sometimes there are problems when fine or sticky bulk material accumulates between the pans.
3. Heavy self-weight.
4. High Capex.

2.8.3 Vibratory Feeders

Vibratory feeders are a simple, efficient, and economical method of conveying free-flowing bulk materials on short distances with a controlled rate.

The principle of vibratory feeder operation is oscillating motion of the trough actuated by a drive that utilizes springs to deliver a linear angle of attack along the entire length of the feeder trough. This oscillating motion causes bulk material, lying on the trough, to move forward making a series of short, rapid hops (Figure 158).

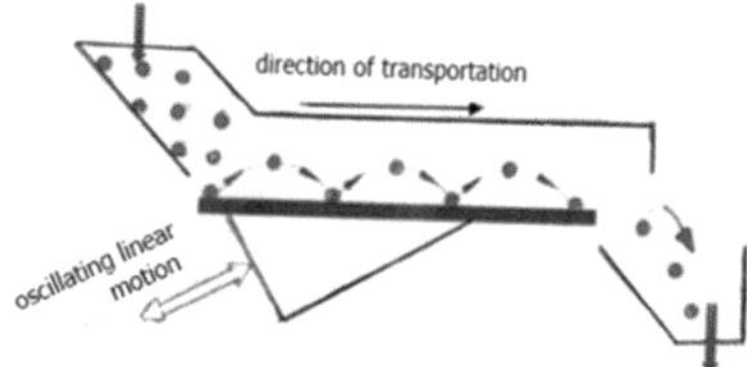

Figure 158. Work principle of vibratory feeder.

Vibratory feeders can be trough, tube or pan types, and can be foot mounted or suspended (Figure 159, Figure 161).

In one of the new projects, we recommended the use of a vibratory feeder to divide the bulk material flow into a number of equal (or unequal, depending on the requirements) sub-flows.

The dividing is carried out precisely because on the outlet of the main (feeding) vibratory feeder, the bulk material moves as a "cake" of equal thickness (or moving layer of rectangular section). So, the width of the fixed receiving chutes determines with great accuracy the portion of the "cake" entering this chute.

The manufacturer of the feeder should supply a main feeder of a length that is sufficient to distribute the bulk material evenly at outlet of this feeder.

For example, a vibratory feeder 3.5 m in length (IFE, Austria) is required to evenly distribute 1,300 tph of granular potash along the pan width on the outlet of the feeder. Figure 163 shows how the falling material proportionally divided into three equal sub-flows by three fixed chutes. Optionally, the material flow can be divided into four or five sub-flows only by the replacement of output chutes.

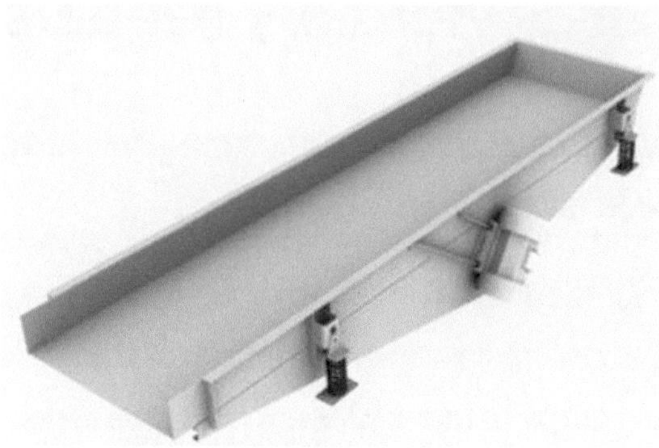

Figure 159. Tray-type vibratory feeder with unbalanced electric motor drive (IFE, Austria).

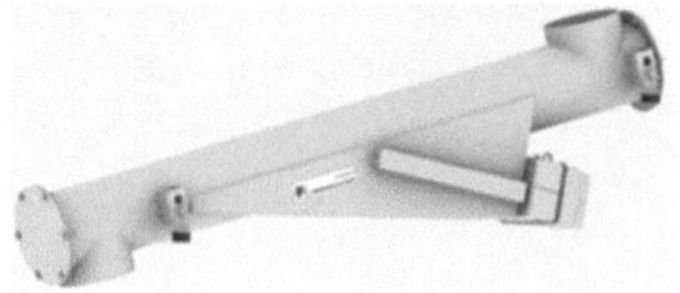

Figure 160. Tube-type vibratory feeder with electromagnetic drive (IFE, Austria).

The vibratory feeder is not intended to be fully installed under the hopper outlet and to be filled with the material, because the oscillating motion of the vibratory feeder and the jumping of the moving layer of particles (Fig. 159) require that most of the pan length will be free of material. The vibratory feeders are usually fed with a chute located in the back of the feeder (Figure 161).

Figure 161. The tube-type vibratory feeder with electromagnetic drive (IFE, Austria).

Figure 162. Vibratory feeders discharge hoppers to feed screeners (IFE, Austria).

Advantages of vibratory feeders:

1. Simple, reliable, and inexpensive, with practically maintenance-free operation.
2. Low power consumption.
3. High capacity that can be easily regulated.
4. No particle attrition due to low material speed (up to 0.3 m/s).
5. No spillage.
6. No dust emission.
7. Possibility of equal (or unequal, if required) separation of the flow of conveyed bulk material into several sub flows by dividing the layer of material at the feeder outlet.

Disadvantages:

1. Maximum conveying length of one feeder is only 10 m.
2. Adhesive, wet, and fine/dusty bulk materials can cause build-up on the tray.
3. High capacities require wide pans (up to 3 m).

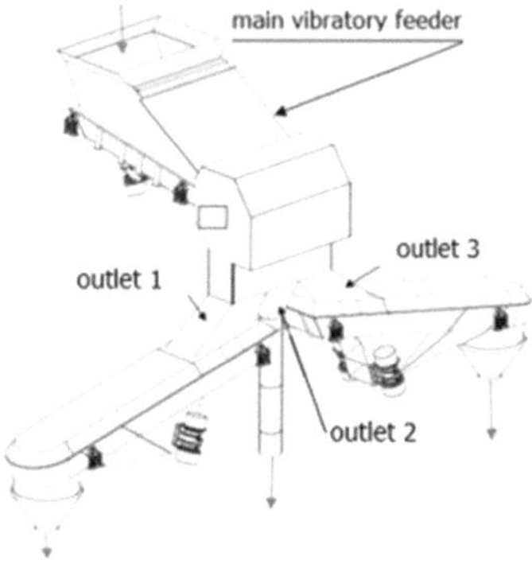

Figure 163. The dividing of 1,300 tph flow of bulk material into three equals sub-flows using two takeaway vibrating feeders (1 and 3) and one discharge chute (2) (IFE, Austria).

2.9. Pneumatic Conveying Systems

General

In principle, pneumatic conveying is the transporting of free-flowing, homogeneous bulk materials through pipelines, using compressed air or vacuum [28, 29, 30].

Pneumatic conveying systems can transport bulk materials through a combination of vertical and horizontal sections of pipeline (connected with 90° bends, with optimal bend (*D /d*) ratio of 10, where *D* indicates bend diameter and *d* indicates pipe diameter). The behaviour of solids in *upwardly inclined* pneumatic pipelines is uncertain and causes clogging of the pipe when the air supply stops suddenly. Hence, inclined pneumatic pipelines should be avoided.

The decision to use (or not to use) pneumatic conveying, and the selection of the right type of pneumatic conveying system, both depend on the characteristics of the bulk material and the length/route of pipeline. To ensure that your choice is the correct one, it is helpful to consider previous experience and the experience of similar facilities or to perform a full-scale test (the optimal solution!) or a reliable bench test.

Pneumatic conveying systems can be divided into two systems that are different in the mode of pneumatic transportation: the dilute phase systems at positive pressure and negative pressure (Figure 164, Figure 165, Figure 166, Figure 167) and the dense phase systems (Figure 164, Figure 168, Figure 169, Figure 170).

The fundamental difference between the two systems lies in the different modes of particle movement inside the pipeline. The difference between the dilute phase and the dense phase modes is shown in Figure 164.

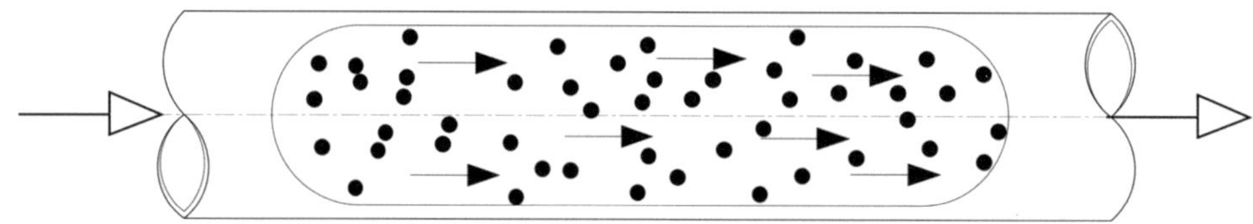

DILUTE PHASE

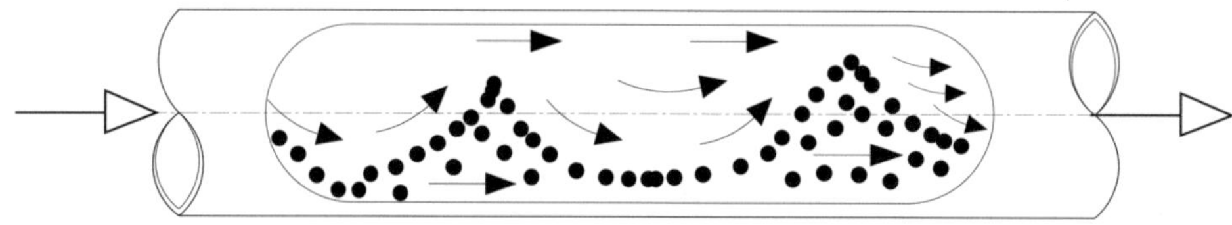

DENSE PHASE

Figure 164. Graphical representation of the two modes of pneumatic conveying. A simplified pneumatic conveying phase diagram is shown in Figure 165.

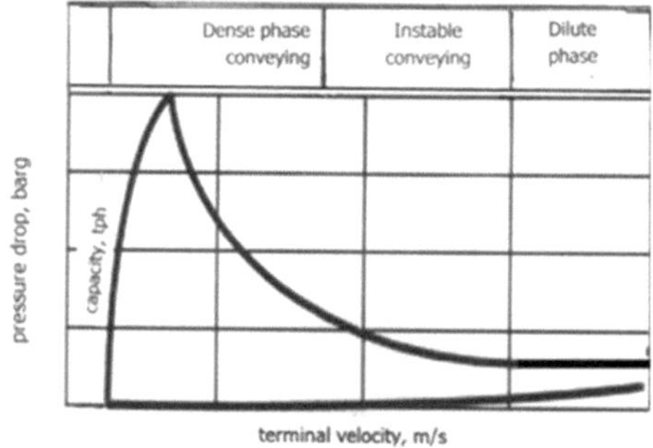

Figure 165. Simplified phase diagram of pneumatic conveying.

2.9.1 Dilute Phase Pneumatic Conveying

2.9.1.1 Positive Pressure Dilute Phase Pneumatic Systems

Dilute phase conveying of dry powders or light granulated particles is a flow of fully separated, suspended particles moving within pipe by high air velocity (from 15 m/s to 40 m/s). The air velocity depends on the dimensions, shapes and specific weights of particles, concentration of particles in the flow and on the length and configuration of the pipeline. In principle, dilute phase conveying is the transportation of separated particles in the by high-velocity airflow (Figure 164) at a ratio of particle velocity to air velocity equal to 0.7÷0.8 for coarse particles and 0.85÷0.95 for fine particles. The systems operate at low pressure drops (1.0 bar to 1.5 bars) and at low solids-to-air ratio (μ_d), where:

$$\mu_d = \frac{\text{bulk material capacity } [kg/sec]}{\text{compressed air capacity } [kg/sec]};$$

μ_d = 15 kg/kg to 5 kg/kg.

The longer transport pipeline, the lower the ratio μ_d and the lower the efficiency of conveying.

Due to the turbulence of the high-velocity airflow, the particles jump from side to side of the pipe. These impacts cause a mechanical breaking/attrition of the particles.

The widely used feeding device for positive-pressure dilute phase pneumatic conveying is a rotary valve which feeds the conveying pipe with a predetermined capacity (to get a solids-to-air ratio required for reliable conveying) and in order to maintain the pressure within the pipeline.

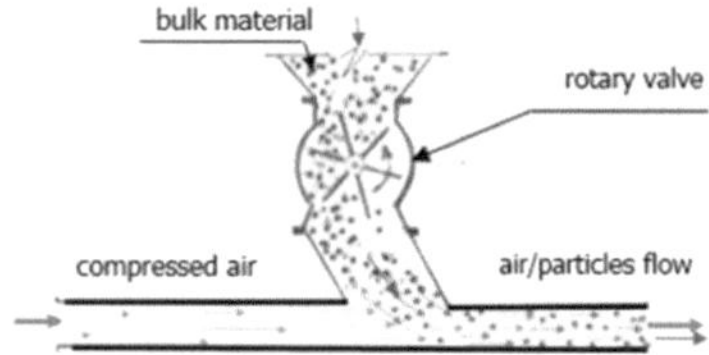

Figure 166. The scheme of rotary valve feeding positive pressure dilute phase pneumatic conveying pipeline.

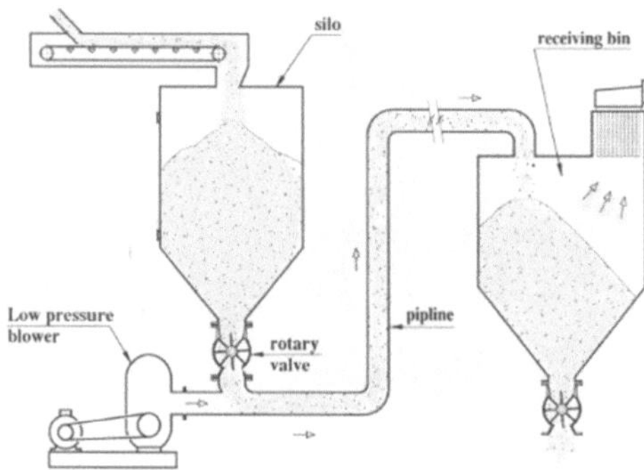

Figure 167. Positive dilute phase pneumatic system fed by rotary valve.

In principle, all calculations of dilute pneumatic conveying are based on Bernoulli's equation for the motion of an incompressible liquid inside a tube between point 1 and point 2:

$$P_1 + \frac{1}{2}\rho v_1^2 + \rho g h_1 = P_2 + \frac{1}{2}\rho v_2^2 + \rho g h_2$$

Where the variables P_1, V_1, h_1 refer to the pressure, speed and height of the fluid at point 1, whereas the variables P_2, V_2, h_2 refer to the pressure, speed and height of fluid at point 2.

A typical positive-pressure dilute phase pneumatic system is shown in Figure 167.

2.9.1.2 Negative-Pressure Dilute Phase Pneumatic Systems

Dilute pneumatic conveying can be carried out also by negative pressure (vacuum).

Vacuum system utilize the vacuum created by a blower or centrifugal fan to pick up and draw in a pipe particles of free-flowing bulk material, to suspend them in the air and thansport the mixture through pipeline to a receiving cyclone/bag filter where the particles are separated from the air and cleaned air is emitted by the fan. The transport vacuum systems usually operate under deep vacuum -4,500 ÷-5,000 dPa.

The vacuum system can be fed by straight suction of the bulk material from an open pile. What is important for such systems is to dilute the bulk material, entering pipeline, to reduce the solids-to-air ratio of conveying to the required for this type of conveying value. For this purpose, special nozzles are used that allow external air to enter the pipe and dilute the mixture (Figure 168).

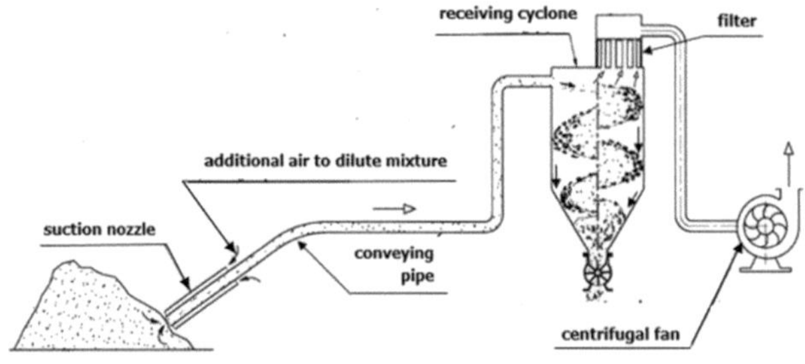

Figure 168. Principle scheme of negative-pressure (vacuum) dilute phase pneumatic system.

The deeper the vacuum the longer the length of transportation by the negative-pressure systems, but the blowers for such systems are much more expensive and more energy-consuming.

The solids-to-air ratio of vacuum systems is varied between 13.0 kg/kg and 3.0 kg/kg, and,

as in positive-pressure dilute phase systems, the longer conveying pipelines, the smaller solids-to-air ratio, the smaller capacity of the conveying and the higher specific power consumption [kW/(t × m)].

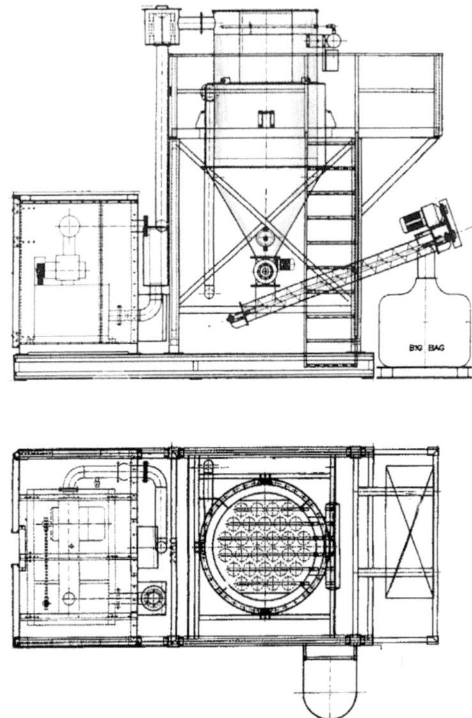

Figure 169. Industrial d ilute phase vacuum pneumatic conveying system.

Figure 169 shows an industrial vacuum pneumatic conveying system that consists of two coaxial cylinders: external cylinder (cyclone) and internal cylinder (bag filters), suction pipe of 120 mm ID, rotary valve, takeaway screw conveyor, blower (operating with a vacuum of minus 4,500 dPa, and air suction capacity of Q = 1,000 Nm³/hr) and compressor for pulse jet bag cleaning.

The "air–particles" mixture enters the gap between the two cylinders tangentially for the primary centrifugal cleaning, and then it cleaned by filtration through the filter bags (secondary cleaning).

The dust is collected in the lower, conical section of the bag filter and evacuated via rotary valve and inclined screw conveyor. The tests on site showed that material capacity of this system was about 10 tph (at length of pipeline about 5 m) at $\mu_d \approx 10$ kg/kg. The pipeline was retracted over 25 m with following drop of material capacity to 3 tph and reduction of solids-to-air ratio to $\mu_d \approx 3.5$ kg/kg.

The vacuum transporter system is reliable and low-maintenance equipment.

2.9.2 Dense Phase Pneumatic Conveying

Typical dense phase pneumatic conveying devices are shown in Figure 170, Figure 171 and Figure 172.

The work principle of dense phase pneumatic conveying:

Unlike dilute phase systems, the transport pipelines of dense phase transport devices (blow tanks - Figure 171, transporters - Figure 172, or airlifts - Figure 174), are fed by the uncontrolled mass of fluidized powders or small granulated particles. Under the effect of pressure drop, the mixture enters the pipeline as a two-phase (compressed air and the fluidized material) flow. Further, on a horizontal section of pipeline, the two-phase mixture is divided into two moving phases: dunes (bottom phase) and airstream (upper phase) (Figure 164, B).

The flow of compressed air (the air velocity is always higher than the dune velocity) passes over the dunes, and creates peaks of air pressure (sharp pressure drops or Venturi throat pressure drops) over each moving dune. Due to this pressure difference, the air saturates the dunes.

The movement of dunes of *powders* (especially easily fluidized powders) saturated with air, is much more effective than the dunes of *granulated* bulk materials because of lower material/tube friction coefficient.

The peak of pressure over each dune creates an additional pushing and moving force on the dune, as the air-saturation reduces the friction of the dune movement. Thus, the capacity and the solids-to-air ratio (efficiency) of conveying are increased significantly. This mechanism can explain why the solids-to-air ratio of dense phase systems (μ_{dp}) is always higher than solids-to-air ratio (μ_d) of dilute phase systems.

The intensive vibration of the horizontal transport pipe, observed during dense phase conveying, is the result of the above-mentioned pulsations of peaks of air pressure passing over dunes.

The mechanism of the air-powders interaction described above, significantly increases efficiency of dense phase pneumatic conveying of such easily fluidized bulk materials as cement, alumina, fuel ash and so on, compared with poorly saturated with air, such as salt, sand or iron powder.

The extension of the length of transportation is accompanied by a further reduction in air pressure and a corresponding increase in actual air velocity. The increased velocity of air destroys the dunes. The destruction is accompanied by a decrease in pressure peaks. The efficiency of conveying is reduced. To keep material in motion, the additional compressed air injection along the pipeline is required, but the injection further increases in the air velocity causes further destruction of the dunes and results in transition to so-called "instable regime". The capacity and the solids-to-air ratio of the conveying are reduced accordingly. Further extension of the pipeline and additional injections of compressed air will inevitably lead to parameters of conveying similar to parameters of dilute phase transportation.

Dense phase conveying operates at high air pressure (usually up to 5.0 bars) and at a relatively low inlet "air–material mixture" superficial velocity of 2.0 m/s to3 .5 m/s, as the terminal air velocity can reach 20 m/s or more. For short distances, the ratio μ_{dp} can reach 200 kg/kg or more, and for long distances the ratio decreases to μ_{dp} = 20 kg/kg÷ 25 kg/kg and less.

Fig. 170 shows a blow tank where an uncontrolled mass of fluidized powders (bottom porous media/septum is used to distribute incoming compressed air to better fluidize the material at the inlet of the pipeline) enters the vertical section of the transporting pipeline.

The conveying of powders over short distances via blow tank is a *self-adjustable mode* where the solids-to-air ratio (μ) has its maximum value for each length of pipeline at the optimal inlet air velocity. The length of conveying pipeline and the superficial air velocity both serve as the specific "feedback" that regulates the density of the two-phase (fluidized bulk material and air bubbles) mixture entering the conveying pipeline. This mechanism was observed in the tests of dense phase transporting via 50 mm ID pipelines up to 40 m equivalent lengths (Figure 171). For longer conveying distances, this mechanism of "self-adjustment" no longer works and the additional compressed air must be injected along the pipeline to continue conveying already in the next, instable mode.

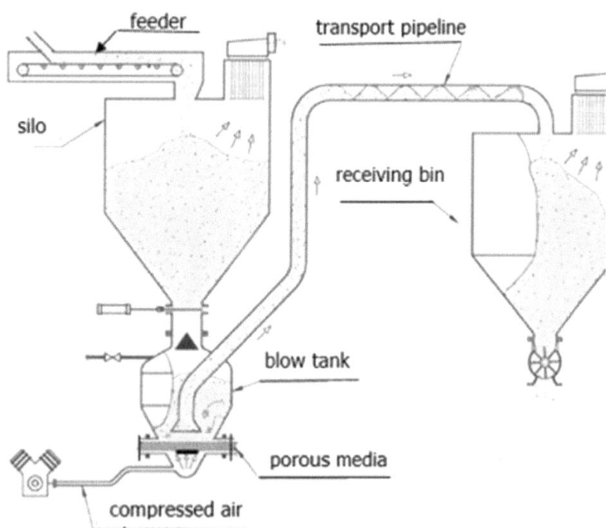

Figure 170. Principle scheme of the blow tank operation.

Figure 171 shows the results of experiments carried out by the author on an experimental rig (a blow tank with a 50 mm ID pipeline; equivalent lengths of pipeline 8.0 m, 16.0 m, and 40.0 m; and the transported bulk material is cement).

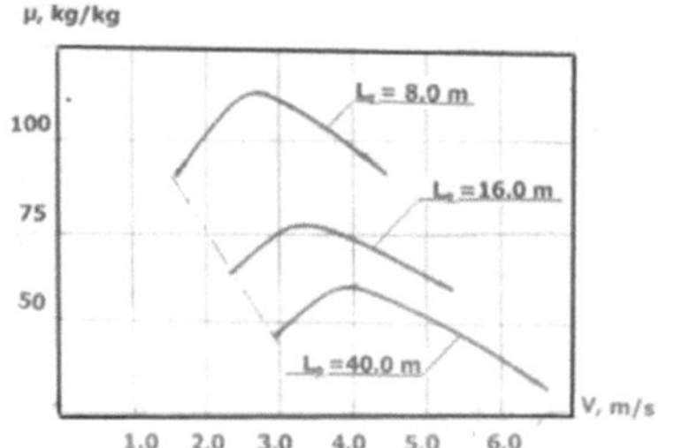

Figure 171. Experimental results of solids-to-air ratios vs superficial air velocities for various conveying lengths (blow tank, cement).

The widely used transporter for dense phase pneumatic conveying is the pressure vessel (Figure 172, Figure 172.1, Figure 173). The pressure vessel differs from a blow tank by the absence of a porous media/septum separating the lower chamber of compressed air from the upper chamber containing bulk material. Instead, the compressed air fed directly into the vessel, at inlet the pipeline, as the pipe is the smooth continuation of the lower conical section of the vessel (Figure 172).

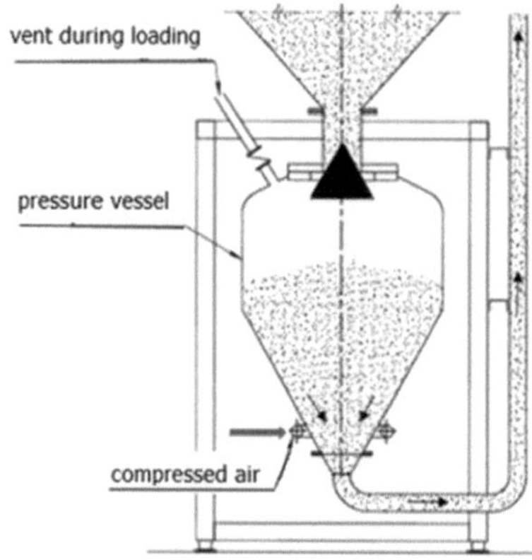

Figure 172. The pressure vessel (transporter).

The high dense air/particle mixture enters the pipeline and must be immediately diluted by additional compressed air injected into the pipeline along the entire length of transporting by special injectors, "air-knives" and so on. This arrangement reduces the solids-to-air ratio but at the same time increases the reliability of the conveying over longer distances.

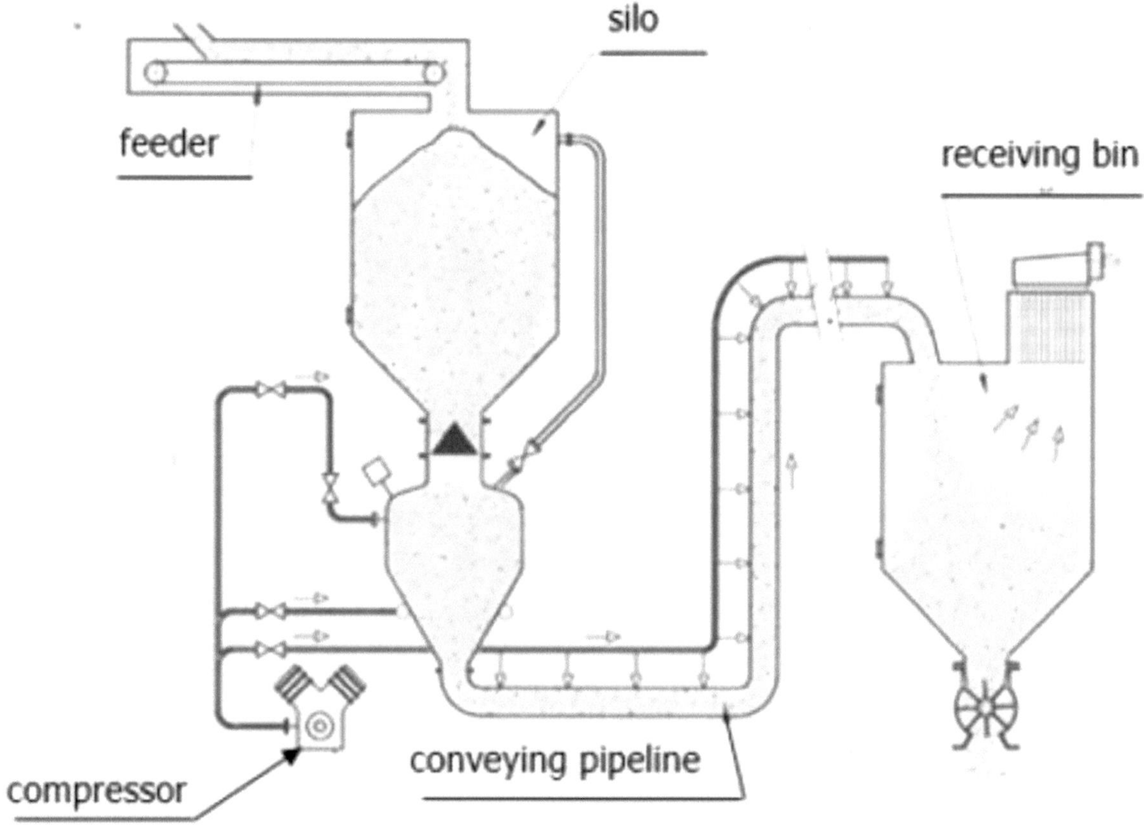

Figure 172.1. Scheme of operation of single pressure vessel/transporter.

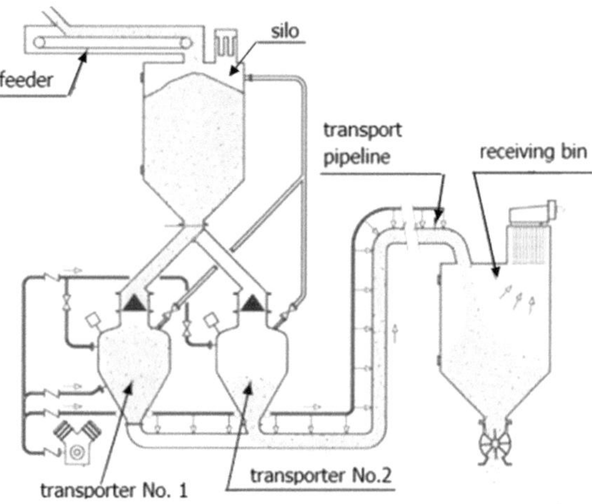

Figure 173. Scheme of operation of dual-pressure vessel/transporter system.

The pneumatic conveying using a single blow tank or pressure vessel is a non-continuous process: the conveying process is stopped each time to fill the vessel from the upper storage silo. The "double- vessel system" overcomes this disadvantage with conveying that is practically continuous: one vessel conveys the bulk material while the second one, in this time, is filled from the upper storage silo (Figure 173).

2.9.2.1 Dense Phase Pneumatic Conveying via Airlift

The air is the simplest pneumatic system, used to elevate powders (cement, flour, apatite, fine phosphate, and so on) with high capacities (up to 1,000 tph) at the height of 10 m to 30 m.

The lower chamber of an airlift (Figure 174) is separated from the upper chamber by a porous media/septum and serves for distribution of air to fluidize powders on the inlet of vertical conveying pipeline. *The airlift operation is based on the fluidization.* Positive air pressure in the upper chamber forces the two-phase mixture (fluidized material and air bubbles) to move toward the inlet of the pipe and enter the pipe. The process is self-adjustable, with the optimal solids-to-air ratio for each given powder, the given air pressure, and the given height of the vertical pipe. The solids-to-air ratio of lifting is varied between $\mu = 250$ kg/kg and 50 kg/kg. The higher the pipeline, the lower the solids-to-air ratio: air expansion and slippage of air bubbles are increased with the height of the pipeline. The airlift does not usually include horizontal pipeline sections where the "material–air" mixture is divided into two phases and requires injection of the additional air. So, the material-to-air ratios of airlifts are extremely high.

One of the principle advantages of an airlift operation is its ability to elevate powders, included relatively large lumps of the same material (Archimedes' principle stating that buoyant forces act on the lumps immersed in the fluidized powders acting as a specific liquid).

In our research (SPB Academy of Sea and River Transport) we reached a capacity of about 900 tph of apatite powder (d_{50} = 60 μm), using 300 mm in diameter and 20 m high pipe at air pressure of 4 bars and superficial air velocity of 3.5 m/s. The optimal solids-to-air ratio of the lifting was 80 kg/kg÷90 kg/kg.

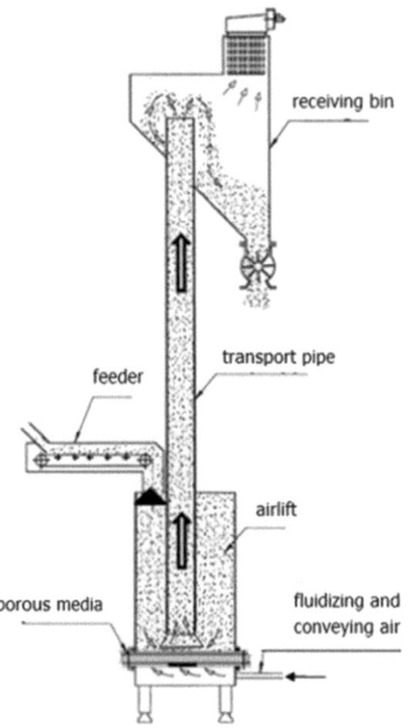

Figure 174. Scheme of operation of airlift transporter.

2.9.2.2 Impulse Dense Phase Pneumatic Conveying

Impulse dense phase pneumatic conveying (developed by the author) was successfully tested in the BMH Laboratory of the SPB Academy of Sea and River Transport in 1970–76 and has been waiting for industrial application.

The *principle* difference between this system and conventional dense phase pneumatic conveying systems consists in replacing of traditional continuous pneumatic conveying with intermittent (alternately stopping and starting) conveying. From the *technical* point of view, the impulse system differs from the conventional dense phase system by the installation of a fast-acting "on–off" valve at the outlet of the conveying pipeline (Figure 175).

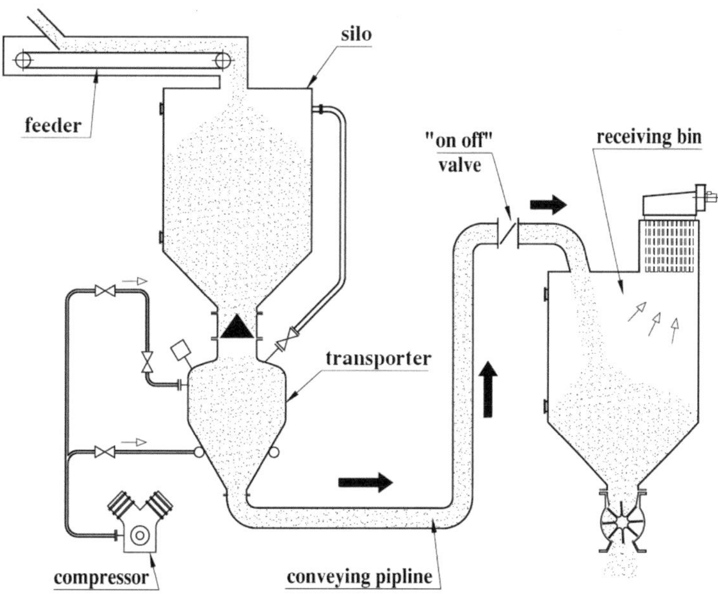

Figure 175. Principle scheme of impulse pneumatic conveying system.

Following are two successive stages of impulse dense phase pneumatic conveying:

1. Static stage

   The "on–off" valve at the outlet of the pipeline (Fig. 175) is closed while compressed air enters "transporter – pipeline" isolated system. The pressure in the system rises to the maximum calculated value, and when the "on–off" valve is rapidly opened, and the transport stage begins.

2. Transport stage

   As the "on–off valve" is opened, bulk material within pipeline is quickly fluidized under a rapid pressure drop (so-called "*flash*" effect) and moved by pressure difference along the direction to the receiving bin, as a homogeneous mass of fine particles saturated with air.

The compressed air supply operates non-stop.

While the material continues to move, the air pressure and, thus, material speed in the system is gradually decreased, and when the lower calculated pressure is reached, the end valve automatically closes.

Since the supply of compressed air does not stop, the pressure in the closed system rises rapidly, reaches the upper limit and the end "on-off" valve opens again to begin a new transport stage.

Note:

The *fast* opening of the end "on-off" valve is critical for the system's normal operation. This effect is like (visually speaking) opening of a bottle of champagne—the quick popping of the cork causes the "*flash*" effect, whereas a slow-motion opening avoids this effect.

What is it –"flash" effect?

The "*flash*" effect of the layer of powder can be explained as the certain state when the *stress* in the powder layer, caused by the rapid (!) expansion of compressed air which is saturated the powder, becomes more than the *strength* of the layer. As a result, the air-saturated powder layer bounces upwards under the effect of the pressure drop.

By the way, this effect can be used also for the effective mixing of different fine and ultrafine powders.

The special quick-open valves, made from corrugated, EPDM rubber (Figure 175.1), were developed by the author and successfully used in the laboratory for the dense phase impulse pneumatic conveying. The rubber insert of this valve is reinforced by fabric layers and can't be stretched. The hermetical closing of the pipe (even when filled with bulk material) takes place by smoothing out of the corrugated insert. Fig. 176 shows that the internal diameter of the rubber insert is larger than the internal diameter of the conveying pipe. This solution results in the reduction of rubber wear and in significant increasing of the service life of the valves.

The valves were tested on pipelines from 20 mm to 300 mm in diameter, and they demonstrated reliable operation and fast opening times (between 0.1 sec and 0.8 sec with a closing pressure difference of ΔP = 0.3 bar to 1.0 bar.

An option: the valve insert can be made from special rubber (e.g. NBR or similar), so the valve can be operated hydraulically instead of compressed air.

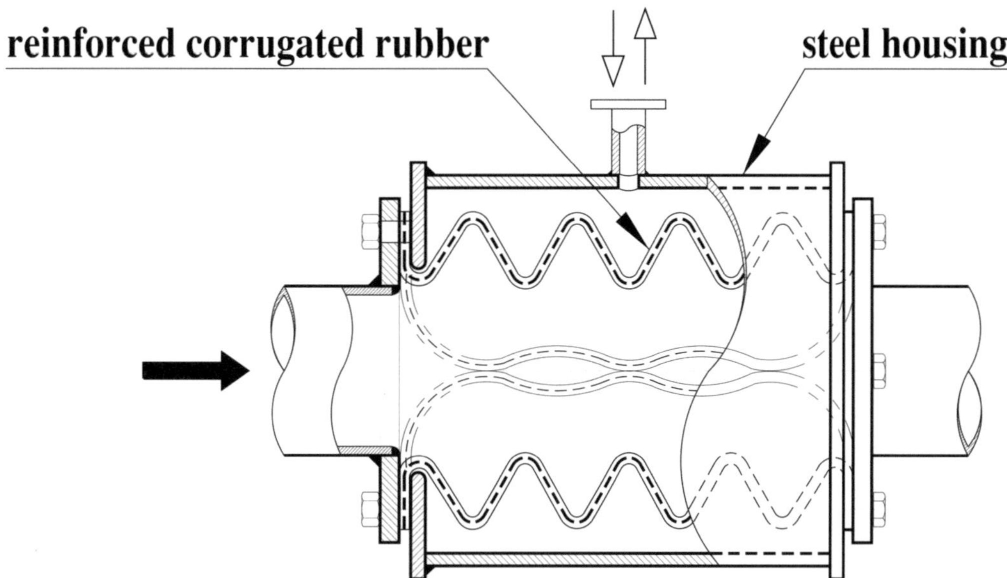

Figure 176. Special "on–off" valve (with corrugated reinforced rubber insert) developed by the author specially for the impulse dense phase pneumatic conveying system.

Advantages of dense phase impulse conveying:

1. Transport process can be stopped and re-started many times (in accordance with the operational requirements) with a pipeline full of material.
2. The average capacity of conveying can be increased by enlarging the diameter of the conveying pipe with a corresponding increase in compressed air supply.
3. There is no blockage of pipeline at all—blockage is one of two stages of the transporting process.
4. The solids-to-air ratio of conveying is higher by 10%–20% compared with the conventional dense phase system.
5. It is possible to distribute the transported bulk material between any number of receiving bins.
6. Increase of air pressure in the system allows to increase the length of the transportation.
7. The system can be also used for conveying of fine powders with very small capacities through very complex pipelines.
8. The bench tester for bulk material, based on this principle, was examined by the MH Laboratory of Ben-Gurion University (Israel) and showed fast, repeatable and reliable results with respect to the dense phase pneumo-transportability of tested bulk materials (see below).

Disadvantages of dense phase impulse conveying:

1. An additional, non-stop operating "on–off" valve should be installed at the outlet of the transport pipeline.

2. Non-continuous transportation.
3. Lack of industrial experience that can show the real merits and shortcomings of this system of pneumatic transportation.

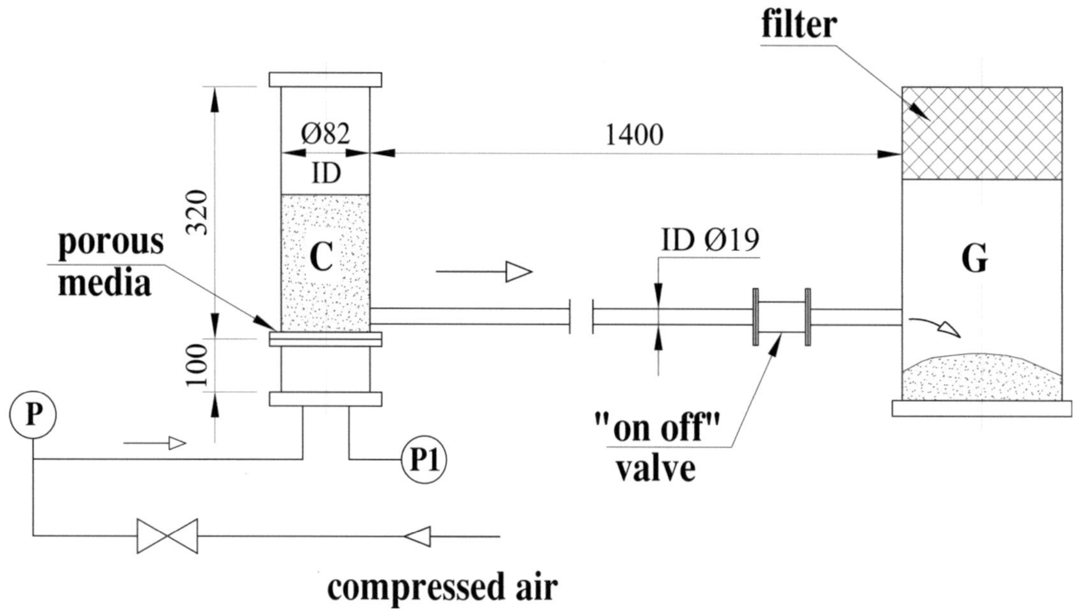

Figure 177. Principle scheme of the dense phase pneumatic bench tester.

This bench tester was developed by the author and was repeatedly used in Russia to recommend (or not to recommend) the tested bulk material for dense phase pneumatic transportation.

The bench tester (Figure 177) consists of two chambers: upper chamber, C, containing a tested bulk material, with a porous septum (e.g., felt) separating the upper chamber from the bottom chamber (air chamber) f; a pipeline (19 mm ID, 1.4 m long); a ball valve on the outlet of the pipe; a receiving chamber, G, with an air filter.

The Test Procedure:

1 kg of the bulk material to be tested is placed into chamber C, and the top cover (blind flange) is bolted. The end ball valve is closed, and compressed air begins to fill the isolated system (the chamber C and the pipeline). When the pressure in the system reaches 1.0 bar, the valve, supplying compressed air, closes. Now, the isolated system (chamber C and the pipeline) is under air pressure of 1.0 bar. Rapid opening of the end ball valve causes the "flash" effect (see above): the bulk material is quickly fluidized and same of the material is ejected by the pressure drop in the chamber G. Weight of the bulk material thrown into the chamber G is the indicator of dense phase pneumo-transportability of the tested material. The more bulk material will be thrown from chamber C into chamber G, the higher the comparative solids-to-air ratio of dense phase pneumatic conveying of the material. For example, about W ≈ 500 g of cement (which is one of the best bulk materials for dense phase transportation) will be thrown into chamber G, in comparison with W≈100 g of granular potash or 10 g of iron powder.

As noted above, the ability to be saturated with air significantly increases the dense phase transportability of the bulk material.

The repeatability of the tests is particularly high if for every new test, the new portion (1 kg) of tested bulk material will be placed in chamber C.

Two or three tests is enough to get a reliable result for the given bulk material.

Now the results obtained on the tester must be compared with the results of testing other known bulk materials (Figure 177.1). For example, we found that the new tested bulk material is very similar to fine phosphate. The next step is to find out where and how operate dense phase pneumatic systems conveying fine phosphate. The parameters of the pneumatic system that the engineer is going to build will be, practically, the same as the parameters of the known systems. The test can show also that the specific tested bulk material *can not* be recommended for dense phase pneumatic conveying and often such a result is more valuable than the theoretical calculations (this has been proved in practice!).

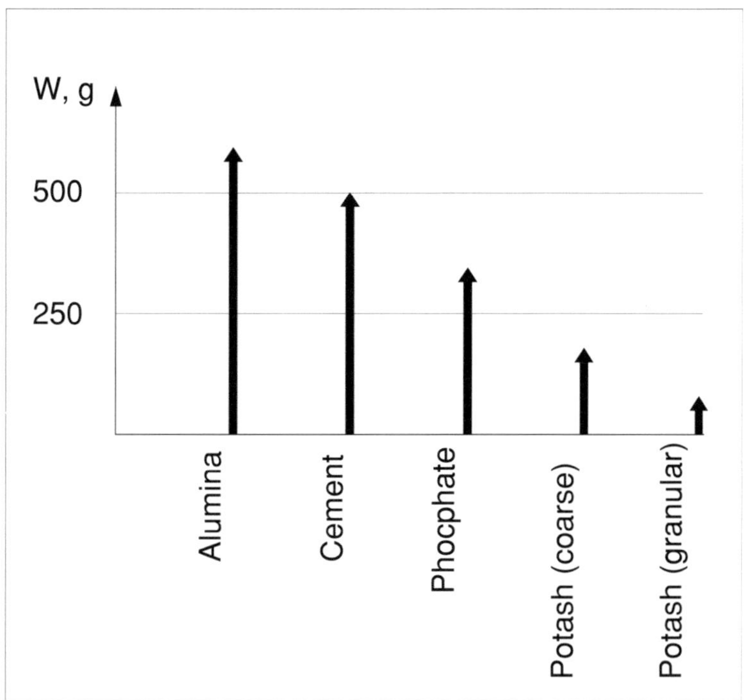

Figure 177.1. Example of bench tests of various bulk materials.

IMPORTANT NOTE:

The generally accepted theory of pneumatic conveying does not exist, as there is no generally accepted method for calculating real bulk material. All recommended theoretical equations, without exception, include correctional empirical coefficients taken from experience—from already operating industrial pneumatic conveying systems, but the coefficients can be used correctly only if the parameters of the new bulk material are *exactly* the same as the parameters of the material transported through the existing pneumatic systems. Objective difficulties lie in the fact that even small changes in particle-size distribution, chemical composition or moisture content of a bulk material can significantly change characteristics of a bulk material and, accordingly, the parameters of pneumatic conveying will be changed.

Conclusions:

At the present stage, only previous experience, or full-scale tests, or reliable bench tests will give you confidence that your new pneumatic conveying system will operate reliably and with required technical parameters.

2.9.3 Special Pneumatic Conveying Systems

- Claudius Peters GmbH (Germany) presented a newly developed pneumatic conveying system, Fluidcon, where easily fluidized bulk material moves through conveying pipe being constantly fluidized. The pipeline is equipped with special fluidizing pads built in the bottom of pipeline. The compressed air feeds the pads installed along the entire length of the pipeline This system differs by an extremely high Capex (the special, complex piping and compressed air injection along the pipeline). We did not find evidence from site that such a system is more effective than other dense phase pneumatic conveying systems.
- Energomar-Nord proposed a pneumatic conveying pipeline with additional compressed air supply through an additional small diameter pipe equipped installed inside the conveying pipe.
- The transhipment floating terminal of Kovako was built for pneumatic unloading of cement, alumina, and similar easily fluidized fine powders. The maximum capacity of the system over relatively short pipelines is about 300 tph.
- Fuller-Kinyon type H pumps, which are special transporters, used for such fluidized powders as cement and fly ash. The material is fed by a short screw conveyor and enters mixing chamber (wind box) where it is picked up by airstream. Then the air–material mixture is directed through a pipeline to the receiving bin.

2.9.4 Pneumatic Conveying vs Mechanical Conveying

Advantages of pneumatic conveying:

1. The conveying of dry powders and light granules (especially if hazardous or dusty) through totally enclosed vertical and horizontal pipelines which corresponds to environmental and safety requirements.

2. Transport pipelines can be easily integrated into existing built-up plant.

3. The high capacity of elevating of the powders by an airlift can be obtained by simply enlarging the diameter of the conveying pipe and corresponding increasing the supply of compressed air.

4. Low- maintenance requirements (if the system was designed correctly and is maintained properly).

5. Construction of the conveying system is relatively simple.

Disadvantages of pneumatic conveying:

1. Only dry, *homogeneous* powders or light granulated bulk materials can be transported confidently by pneumatic systems through relatively short pipelines consisting of vertical and horizontal sections.

2. It is desirable to screen the material prior to conveying to prevent accidental penetration of foreign objects or oversize lumps into pipeline usually ending with pipe clogging and stopping of conveying. To re-start the conveying, the pipeline should be taken apart and cleaned manually.

3. Pneumatic systems have the highest power consumption [kW / (t x m)] in comparison to other conveying systems (Table 2, section 2.1.1). The major reason for this cost is the excessive direct expenses for compressed air supply due to ineffectiveness of air-compressing system. The overall efficiency of a typical air-compressing system can be as low as 10% to 15%. For example, the cost of compressed air (8 bars) for a standard 37 kWh compressor is about 0.016 $/m^3 and the annual (6,000 work hours) electricity cost of 1.0 CFM (or 1.685 m^3 /hr) is about $100.

4. Pneumatic conveying shows a sharp decline in capacity and in the solids-to-air ratio (efficiency) with increase of the conveying distance.

5. Transporting of abrasive powders with high-velocity airflow causes rapid wear of the pipeline walls.

6. The expensive air separation and dedusting systems must be installed on the outlet of the pipeline to satisfy clean air requirements.

7. The combined "vacuum-in and air pressure-out" conveying systems (Fig. 178) decrease the total efficiency of conveying. The positive pressure at the outlet of the pipeline reduces the vacuum/ suction capacity at the pipeline inlet.

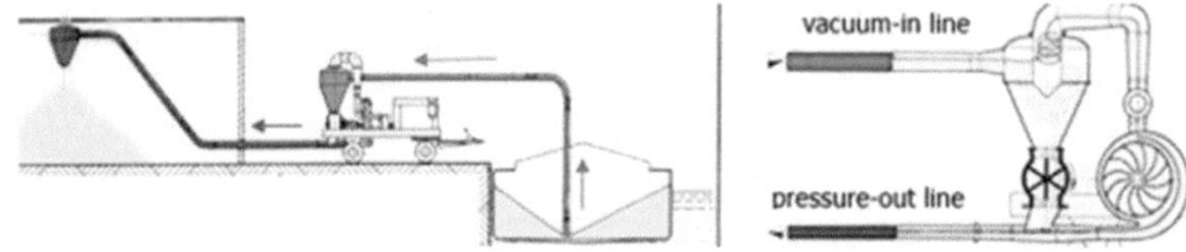

Fig. 178 "Vacuum-in and pressure-out" pneumatic conveying system (Neuero Corp.).

2.9.5 Capsule Pneumatic Conveying System

Capsule pneumatic conveying system consists of a fixed steel pipeline and a container/s (capsule/s), with wheels at each side for guidance and support, that are moved by the air pressure difference along the pipe as capsule seal prevents/reduces airflow between the capsule body and the pipe walls (Fig. 179).

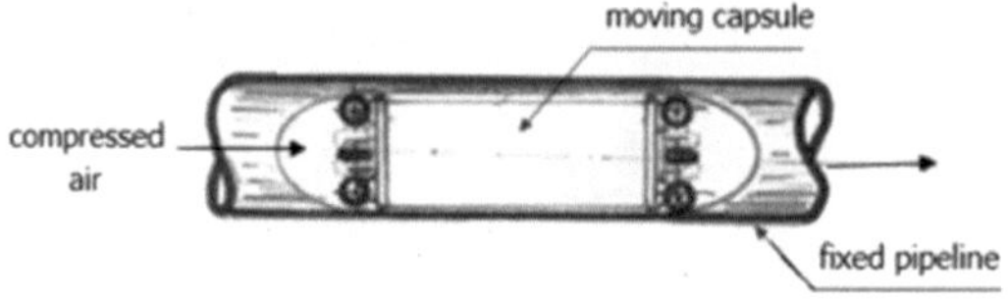

Fig. 179 Principle scheme of the transport capsule.

The radius of the horizontal curve of a capsule route is:

$$R_{min} = 30 \times D,$$

where *D* is the pipe diameter, *m*.

An existing operating system has a pipe diameter of 500 mm and a capsule 2.5 m in length and carries about 500 kg of bulk material at a specific power consumption of about 0.6 kW per km ton/hr.

For comparison, the specific power consumption of the Sdom–Tsefa Cable Belt overland conveyor (18.2 km long and a lift of about 800 m, at capacity of 650 tph) is only 0.25 kW per km x ton/hr.

Without positive technical data from the site it is difficult to recommend this system for bulk material handling, but.... it's possible that in the future such capsule transport as HYPERLOOP will replace the well-known today overland conveying systems.

CHAPTER 3.

Choosing of Covered Storage for Bulk Materials

General

Covered storages for bulk materials (Dome-type, rectangular storages, silos, bins, and so on) are one of the important components of any material handling terminal, plant, seaport or riverport. Such bulk materials as coal, cement, grain, fertilizer, animal feed and so on, must be kept covered to prevent material degradation from winds, rains and so on, and to eliminate dust emission in accordance with very strict local and international environmental rules and regulations (ISO 14001 EMS and others).

The type, the main dimensions of, and the related material handling systems of the storage must be selected after a comprehensive analysis of such factors as required storing capacity, physical space available, bulk material properties and moisture content, loading and unloading rates, ambient climate conditions, duration of storing, and environmental requirements.

The loading and unloading equipment should be reliable, low-maintenance, and should provide the required capacity.

One of the essential characteristics of any bulk material is its tendency to agglomerate or consolidate when the bulk material loses its free-flowing ability (Figure 180). It should be considered together with any long-term storing (weeks or months), moisture content of the bulk material, and environmental relative humidity.

These variables are directly affecting the choice of the optimal material handling equipment.

Figure 180. Bulk material, agglomerated in a storage.

Specific requirements are made for storing flammable bulk materials. The Standard for the Fire Protection of Storage (NFPA 230, 2003) makes recommendations related only to special flammable products (some types of coal, pyrophoric metals, solid waste, rags, paper, wood, dusty and ignitable fibres and so on.).

3.1 General Requirements for Choosing the Type of Storage

Bulk storages, especially port bulk storages, are the buffers between incoming (by train, by truck, and/or by ship) and outgoing (by bulk vessel and/or barge, by train, or by truck) bulk material streams (Figure 181).

Figure 181. Example of port bulk handling facility.

3.1.1 The Choice of Port Storage Capacity

Based on statistics, the required capacity of bulk storage in a port is between 4% and 5% of the terminal annual throughput capacity/freight turnover. The storage capacity can be reduced to 3%, but this would require an increased level of operational planning not only to ensure sufficient accumulation of bulk material for continuous shiploading operations but also to maintain the store capacity at a level high enough for the concurrent discharge of trains or trucks. The large vessels (Panamax type and larger) and the number of different grade of bulk materials that need to be stored separately, require a larger storage capacity which can reach 10% of the freight turnover.

There are a large number of different storages for bulk materials and some of them are systematized below:

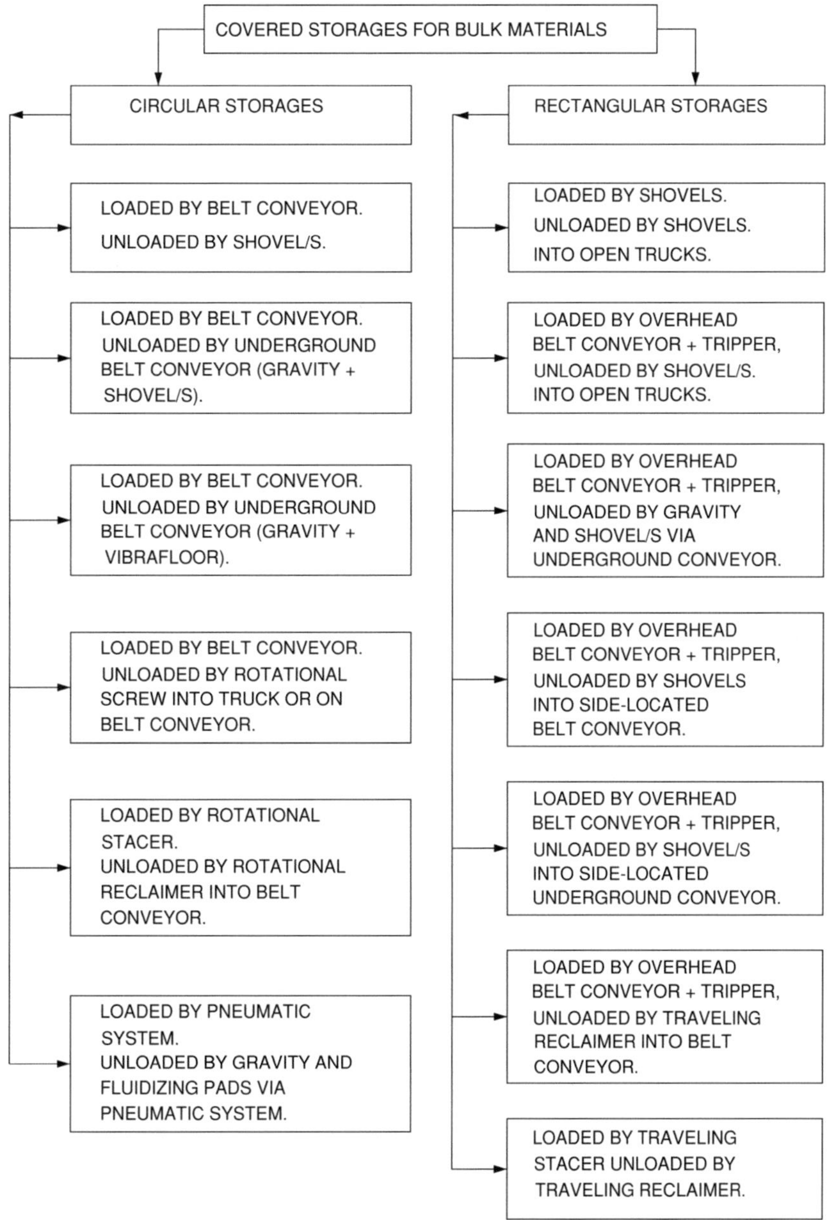

The choice of the optimal type of storage is based on previous experience and on an analysis of existing storages.

3.2 Circular Storages

3.2.1 Storages Loaded by Belt Conveyor and Unloaded by Shovel

Such storages usually have relatively small store capacities (up to 2,500 tons) and are intended for storing fertilizers, de-icing salt, and so on. The storages are used by local distributors, so there is no need for continuous loading operations and for large discharge capacities.

The storages are usually loaded with inclined belt conveyors, that are fed by tipping trucks or shovels. The trucks are unloaded into open shallow discharge pits, located outside of the storages. The head section of the loading conveyor is usually suspended to the centre of the storage roof. The storage is unloaded by a shovel entering the storage through a gate, digging the bulk material from the pile, and pouring it out into takeaway trucks waiting outside.

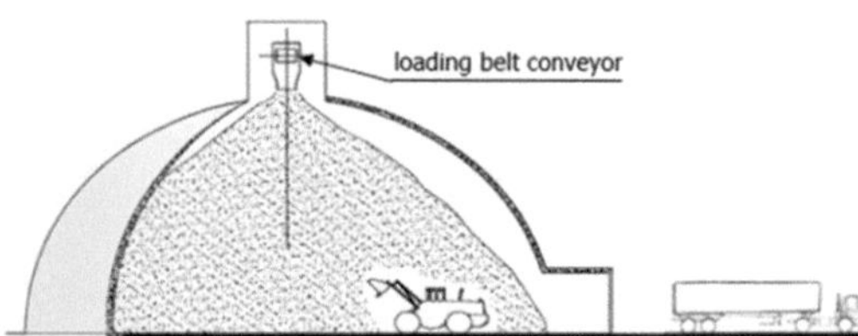

Figure 182. Circular Dome-type storage loaded by belt conveyor and unloaded by shovel into trucks.

3.2.2 Storages Loaded by Belt Conveyor and Unloaded via Underground Belt Conveyors

We will begin reviewing such storages by analyzing one of the existing storages. The existing circular Dome-type storage, shown in Figure 183, was designed and built for bulk potash in 1974–75 and successfully operates today.

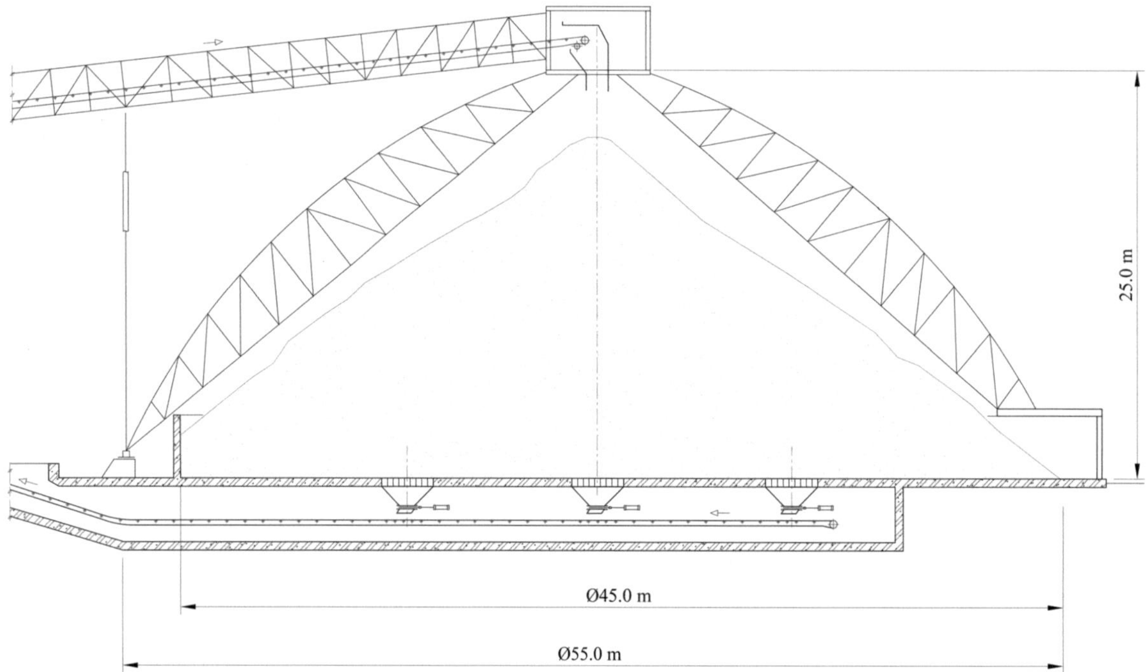

Figure 183. Circular Dome-type storage.

The ratio of the storage height to its internal diameter is approximately H/D=0.55 (Figure 183). The concrete retaining walls of the storage are 4 m high, and the storing capacity of the storage is 15,000 t to 16,000 t of bulk potash. The storage is loaded by inclined outer conveyor through an opening of the centre roof and is unloaded by an underground belt conveyor fed through three underground hoppers, 11 m between them. In the original design, each hopper was equipped with a belt feeder to transfer potash from the hoppers to the underground conveyor. Because of heavy maintenance problems, these feeders were dismounted in the initial stages of the storage operations and were replaced by chutes with slide-type gates that proved to be the simplest, low-maintenance and the most reliable equipment.

Advantages and disadvantages of this existing circular Dome-type storage are noted below.

Advantages of the existing storage:

1. The storage is loaded via one central loading opening, straight from the head chute of the inclined feeding conveyor, so an overhead conveyor with a tripper is not required.
2. About 30%÷50% of the stored free-flowing bulk material is discharged by gravity flow and the remaining material is relocated to underground hoppers by shovel.
3. The existing problem of dust emission from the open storage gates during loading and unloading of dusty bulk materials can be solved using a dedusting system with a number of suction hoods installed around the gates, as it was carried out in 2012 on the gates of the phosphate storage in port of Eilat (Figure 184). This dedusting system has been operating with no dust emission from gates that open during loading and unloading of dusty phosphate.
4. Agglomerated, consolidated materials can be discharged (of course, with less capacities, but discharged!) by shovels.
5. As an option, a telescopic Cleveland Cascade chute can be suspended to the chute of loading conveyor. The Cleveland chute significantly reduces speed of fall of the bulk material and thus reduce particle attrition and dust generation, but this solution has several serious disadvantages that will be discussed later (see section 3.3.3.4).
6. The Capex is relatively low.

Disadvantages of the existing storage:

1. The first discharge underground hopper is located at about 12 m inward from the gate, so when the storage is fully loaded, the bulk material covers the opening of the first hopper, preventing the shovel approach to the hopper. From the beginning, the shovel should take some material out of the storage and pile it outside to ensure an access to the first underground hopper in order to begin loading the underground conveyor. To prevent this time-consuming operation, the operator must prevent the fully loading of the storage keeping free approach to the first hopper for the shovel.

2. In the case of non-free-flowing material to be stored, the discharge efficiency is reduced: only one shovel can be used in such storage, at a maximum discharge capacity of about 650 tph.

3. Loading of this storage by the overhead belt conveyor concurrently with shovel discharge operations can cause a serious safety problem for shovel operator (heavy dust inside of the storage).

4. Only one type/grade of bulk material can be stored.

Figure 184. Dedusting of storage gate (the storage contains dusty bulk material).

The difference between these two types of circular Dome-type storages, shown in Figure 183 and Figure 185, is the cupola/Dome construction.

For the conventional storages, the cupola is built as a number of interconnected steel or concrete beams covered by PVC plates or similar lining, while for the new generation of Dome storages the cupola is built using concrete grout injection (shotcrete) on the specially prepared multi-layered semi-spherical roof of the storage.

Most of the Dome-type storages have flat concrete floors, ground-level entrance gates, and discharged through underground hoppers feeding takeaway underground belt conveyors (30%–50% of the free-flowing material is discharged by gravity flow, and the remaining material is discharged by shovels).

The height-to-diameter ratio of most of such storages is $H/D \cong 0.5$.

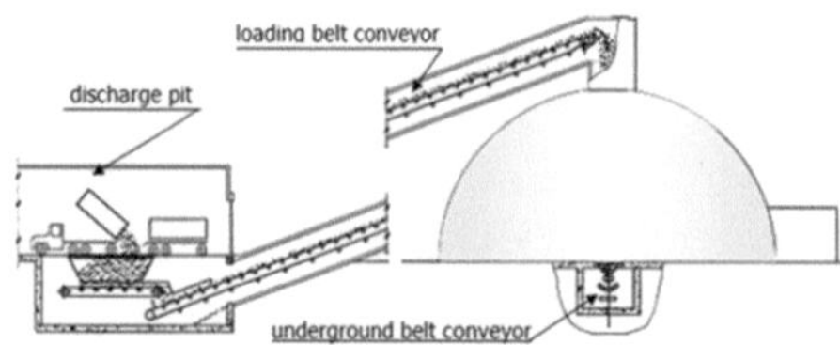

Figure 185. Dome-type storage loaded by inclined belt conveyor and unloaded by shovel and underground belt conveyor.

3.2.3 Circular Dome-type Storages Loaded by Belt Conveyor and Unloaded via Underground Belt Conveyor with the Help of Vibrafloor

Circular Dome-type storages are usually designed with a height-to-diameter ratio of $H/D \approx 0.5$, although there are many exceptions to this rule. For example, two Dome storages (each 36 m in diameter, 34.1 m height, and 36,000-ton capacity) were designed for one of the European ports in 2013 (Figure 186).

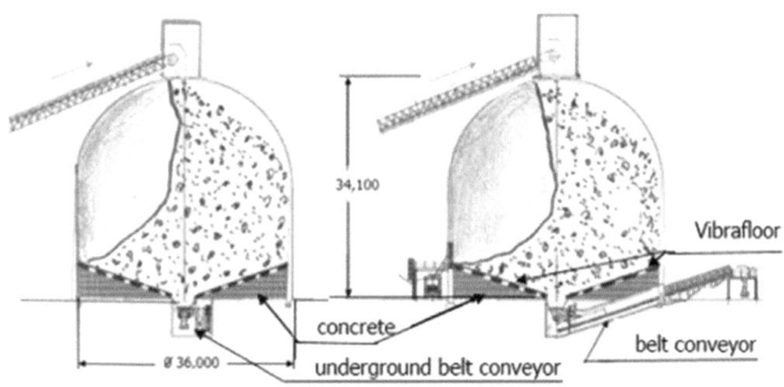

Figure 186. Design of Dome storages (36,000 t each) with Vibrafloor.

These storages were designed to be loaded by inclined belt conveyors through openings in the centre of the storage rooves. The project includes Vibrafloor discharge of the storages: the concrete floors of the storages were slightly declined (12°÷15°) toward the centre line, where were located underground hoppers feeding the underground discharge belt conveyor (Figure 186). The floors were designed especially for Vibrafloor installation.

The Vibrafloor is composed of a number of flat plates installed on the slightly declined concrete floor and overlapping each other. Each plate rests on short flexible supports, and equipped with an electrical vibrator protruding upwards in the centre of the plate. According to the design concept, the vibrated plates must activate all bulk material and direct it to the discharge hoppers. The idea of the project engineers was to use the vibration of the plates to *force all* stored bulk material to move towards the underground discharge conveyor and to unload *all* stored material without shovels.

The storages were designed to store bulk fertilizers in a seaport.

The operation of storages with Vibrafloor is based on one condition: *the bulk material must be free-flowing, dry and light (seed grain, wood chips, so on)* because most of the stored material (85%÷90%) must be discharged by gravity and only remaining material can be activated and moved towards discharge hoppers by the Vibrafloor. If the bulk material is the non-free-flowing, the insignificant capacity of the Vibrafloor vibrating plates (about 40 W/m^2) will not be sufficient to activate and move the large masses of agglomerated or consolidated bulk material towards the underground hoppers. In this case, the discharge process will stop.

The release such a storage from consolidated bulk material, with no access for shovels, will be an impossible mission even for the most experienced terminal manager.

After a detailed technical analysis, taking into consideration that fertilizers are non-free-flowing bulk materials, especially in the humid environment of a seaport, this project, completely designed and having passed the tender process, was rejected.

3.2.4 Circular Silo Loaded by Belt Conveyor and Unloaded by Rotational Screw

The Eurosilo concept (Figure 187) is an attempt to create a new, progressive technical solution for automatically loading and automatically unloading large circular silos for bulk materials.

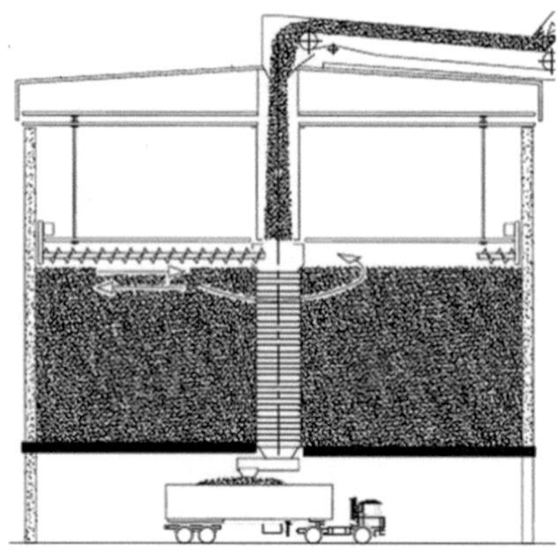

Figure 187. The scheme of Eurosilo.

For example, ESI Eurosilo Co., that already designed the 50,000 m³ silo, currently works on a project of two 100,000 m³ silos (55 m in diameter and 42.5 m in height) for steam coal. Each silo is equipped with a rotating screw conveyor and a central slotted column which ensures reliable discharge of sticky materials. The screw is suspended to a rotating, and moving up and down bridge structure.

During loading of the silo, the rotating screw evenly distributes the bulk material in the silo. During discharge, the screw, rotating in the reverse direction, moves the stored bulk material inwards, through the slots in the central column. The material falls freely and loads the trucks (straight or by belt feeder) or a takeaway belt conveyor.

A similar loading–unloading system (rotational screw mounted on the central column) is used in circular Dome-type storages ($H/D \approx 1.0$) manufactured by Domtec International.

Advantages of the Eurosilo:

1. Minimal footprints for large storages.
2. Relatively low degradation of material (coal, petroleum coke, crushed limestone, etc.).
3. Automatic concurrent filling and reclaiming of bulk material with controlled capacity.
4. No shovels are required to discharge the storage,

Disadvantages of the Eurosilo:

1. The complex loading and unloading mechanisms that are constantly inside of the silo, in a dusty, often corrosive environment, require regular inspection and maintenance, resulting in high Opex.
2. The discharge of agglomerating and consolidating bulk material, especially in the case of long storing periods, becomes a serious problem. This is the reason why the storages are used only for non-consolidating bulk materials (coal and similar materials).
3. Only one type/grade of bulk material can be stored.
4. High Capex.

3.2.5 Circular Dome-type Storage Loaded by Rotational Stacker and Unloaded by Rotational Reclaimer

The automatic loading and unloading of bulk materials (e.g. coal, ore) is shown in Figure 188, where the circular Dome-type storage is equipped with a swivel loading conveyor (stacker) and with a swivel scraper reclaimer fixed in the centre of the storage. The reclaimer transfers the bulk material to the central underground hopper feeding underground takeaway conveyor. The stacker piles the material around in circular piles and can operate concurrently with the reclaimer.

Figure 188. Circular storage with fixed swivel stacker and swivel reclaimer (Bedeschi, Italy).

Advantages of this circular Dome-type storage:

1. Fully automatic loading and unloading.
2. Controlled unloading capacity.
3. The loading and unloading can be carried out concurrently.
4. No shovels are required for the loading and unloading operations.

Disadvantages of this circular dome-type storage:

1. The complex loading/unloading equipment, located in a dusty and often corrosive environment, requires regular inspection, monitoring, and maintenance carried out by a skilled maintenance team.
2. High Opex and high Capex.

3.2.6 Circular Dome-type Storage Loaded and Unloaded by Pneumatic Conveyor

The circular Dome-type storage shown in Fig. 189 contains cement (or similar fine, easily- fluidized powders) which is loaded and unloaded by pneumatic conveying systems. Most of the material is discharged by gravity through dense phase pneumatic systems. At the final stage of the discharge, the compressed air actuates fluidizing pads, installed on the declined concrete floor, fluidizes, moves and discharges the remaining cement that was not discharged by gravity.

Figure 189. Dome-type storage for cement (H/D ≅ 1) that is loaded and unloaded by pneumatic conveying systems (Domtec International).

3.3 Rectangular Storages

Rectangular covered storages can be loaded by shovel, by stacker, or by overhead conveyor with tripper, and unloaded by shovels, by side-located belt conveyors, or by reclaimers.

3.3.1 Rectangular Storages Loaded and Unloaded by Shovel

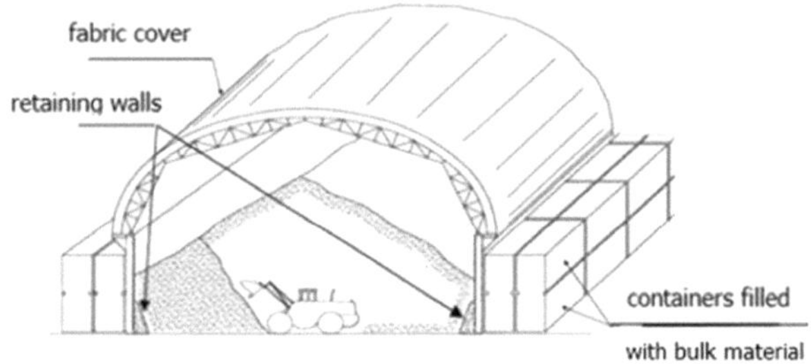

Figure 190. Fabric-covered rectangular storage, loaded and unloaded by shovel.

Figure 190 shows a low-cost fabric-covered rectangular storage that loaded and unloaded by a shovel.

The cover of the storage is supported by the upper structure consisting of steel arches. The walls of the storage are supported by the standard containers filled up with a bulk material.

The prefabricated concrete retaining walls of about 2.0 m high are installed along the length of the storage in order to prevent damage caused by the shovel operation.

The structure can be covered with UDWC (ultradurable woven canvas).

According to an FAO (Food and Agriculture Organization, UN) corporate document (1992), the plastic covers for bulk storages may cost as much as USD$2 per stored ton of bulk material, depending on the material used.

Heavy-duty PVC covers can be expected to last three or more seasons (depending on the rate of UV degradation), while lighter-weight polyethylene covers are best to be replaced every season.

Advantages of fabric-covered rectangular storages:

1. Quick design and short construction period.
2. Relatively high Opex.
3. Low Capex.

Disadvantages of fabric-covered rectangular storages:

1. Small storage capacity.
2. The storage must be built with concrete floor.
3. Small loading and unloading capacities with shovels.
4. Not recommended for dusty bulk materials due to dust, generated during loading and unloading by shovel/s, as high dust emissions cause severe ecological problems.
5. Fabric cover is usually replaced every three or four years.

The hard-roof multi-bay storage with concrete floor and walls that loaded and unloaded by shovels, is shown in Figure 191. The bays are located along two sides of the storage and partitioned by concrete separation walls. The central passageway is wide enough to enable shovel operations. Each type of bulk material is piled up within dedicated bay and discharged by shovel into trucks waiting outside.

Figure 191. Multi-bay storage with wide central passageway.

Figure 192. Covered storage with concrete retaining walls, loaded and unloaded by shovels.

The storage (Figure 192) was designed to store only one grade of bulk material, that is piled up and unloaded by a shovel. The self-made extension boom (shown in Figure 192) is attached to the shovel to ease piling up the bulk material.

The loading of the storage by shovel minimizes Capex and maximizes Opex. The piling of high piles by shovel is a difficult and unsafe operation, especially when the material is dusty. The capacities of shovels operating in such storages are relatively low.

3.3.2 Rectangular Storages Loaded by Overhead Belt Conveyor with Tripper and Unloaded by Shovel

The Dome-type rectangular storage with a hard roof, loaded by overhead belt conveyor with a tripper moving (or moved) along the storage to distribute the bulk material, and unloaded by shovel, widely used in various industrial areas (Figure 193). These storages usually store one type of bulk material, but they can be divided with separation walls to store a few types (or grades) of bulk materials (Figure 194).

Figure 193. Typical covered rectangular storage for bulk materials.

Figure 194. Covered rectangular storage loaded by overhead conveyor with tripper and unloaded by a shovel.

3.3.3 Rectangular Storages Loaded by Overhead Belt Conveyor with Tripper and Unloaded via Underground Belt Conveyor

Rectangular storages loaded by overhead conveyors with trippers and unloaded through underground belt conveyors, are the most widely used storing facilities in seaports. The main advantages of these storages are the simplicity, reliability, high loading capacity, and high unloading (shiploading) capacity. No matter whether a bulk material will be agglomerated or consolidated, it can be discharged by shovels.

The storages are usually unloaded through underground longitudinal conveyors fed through a number of underground hoppers with grizzly-type covers. Depending on flowability of the bulk material, between 30% and 50% of the stored material can be discharged by gravity, and the remaining material is transferred by shovels into underground hoppers.

Gravity discharge of free-flowing material provides maximum discharge capacity and uniform, stable feed of the discharge belt conveyor and this is especially important for the shiploading.

The hoppers are located along the centre line of the storage with a distance of 6 m to 10 m between them (Figure 195, Figure 196). The discharge capacity of shovels usually much smaller than that of gravitational discharge (especially when a shovel must drive a long distance between piles and discharge hoppers), so a width of the storage more than 40 m÷50 m is not efficient.

Different bulk materials can be stored in one storage, where they are divided by special separation walls.

A typical rectangular storage with discharge openings in the floor, an underground gallery with discharge belt conveyor is shown in Figure 195, Figure 196, Figure 197, and Figure 198.

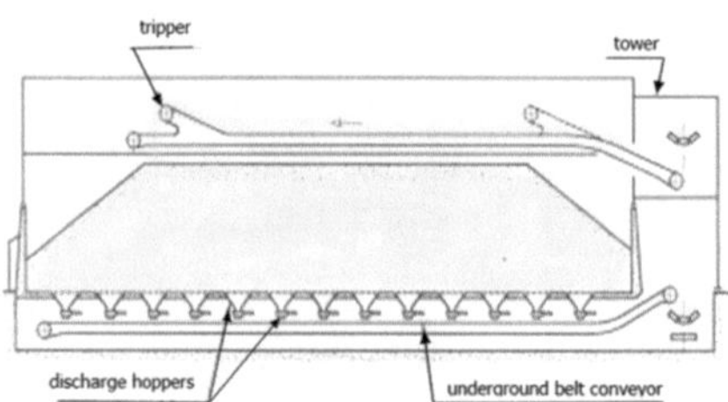

Figure 195. Typical rectangular storage.

The openings in the storage floor with grizzlies (gratings) are shown in Figure 200. Discharge of a storage through openings that not covered by grizzlies is prohibited for safety reasons. The maximum recommended grizzly openings are 110 mm × 200 mm.

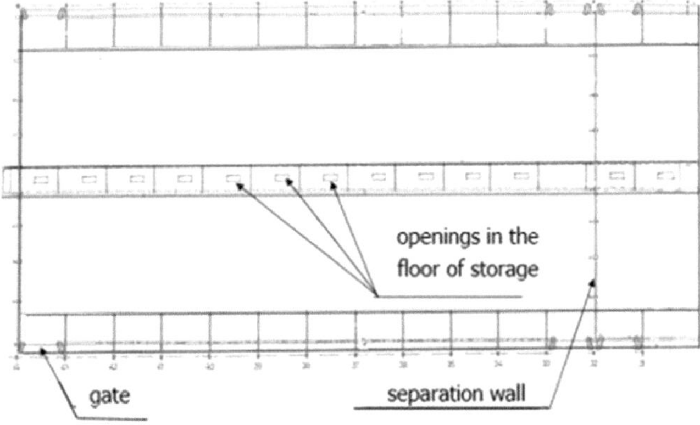

Figure 196. Layout of floor level of a storage with discharge openings and separation wall.

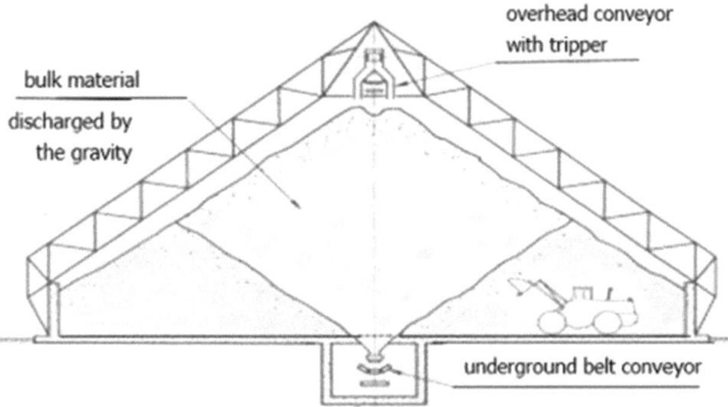

Figure 197. Cross section of a typical rectangular storage.

Figure 198. Typical discharge opening with grizzly cover.

In the early seventies, in one of the northern Russian ports was built a rectangular storage that was slightly different from conventional storages (Figure 199).

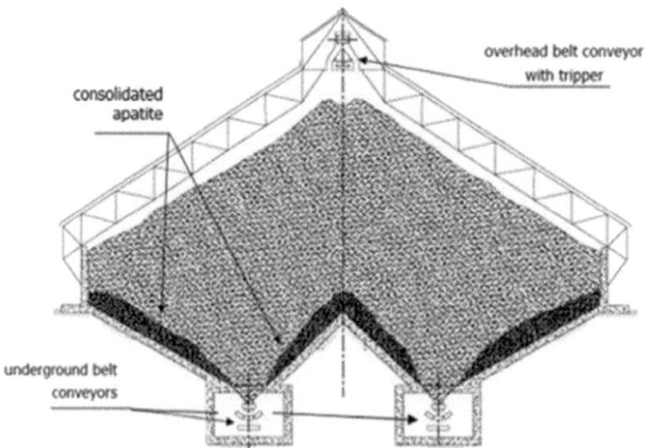

Figure 199. Storage for apatite, unloaded by two underground belt conveyors.

The idea of the project team was to build a new type of covered storage that would be loaded by overhead belt conveyor with a tripper and unloaded by gravity only through two parallel underground conveyors.

Considering that apatite is a dry, free-flowing, easily fluidized powder (d_{50} ≅ 60 μm), the project engineers supposed that after most of the apatite was discharged by gravity, the rest of it would be fluidized (after being released from the weight of the upper mass of material) and would freely slip to the opening of the hoppers. In this way, the storage would be fully unloaded without shovels.

The reality differed from the design considerations: the fine particles amassed on the decline surfaces (especially during long waiting periods between vessel approaches), and in the end, the consolidated apatite covered and clogged the openings into the discharge hoppers (Figure 199). The operation of the storage was ceased. Only after manual cleaning of the openings, the storage operation could be restarted. Since then the manual cleaning continues on a regular basis.

The main components of rectangular storages that will be considered below are:

- 3.3.3.1 Underground belt conveyors
- 3.3.3.2 Gates between underground hoppers and discharge conveyors
- 3.3.3.3 Separation walls
- 3.3.3.4 Trippers and tilting probes

3.3.3.1 Underground Belt Conveyors

Depending on the type of conveyed bulk material and the local safety and environmental requirements, underground discharge conveyors can be open type (Figure 200), guarded (Figure 201), or can be enclosed along entire length of the conveyor in order to prevent dust emission (Figure 202).

All the safety rules noted in section 2.1.3 must be realized for underground belt conveyors to prevent accidents.

Two-side walkways (600 mm to 800 mm wide) make cleaning, inspection, and maintenance operations much easier.

All maintenance works must be carried out on the stopped underground belt conveyors.

Figure 200. Underground gallery with an open discharge belt conveyor.

Figure 201. Underground gallery with a guarded discharge belt conveyor.

Figure 202. Underground gallery with enclosed discharge belt conveyor.

3.3.3.2 Gates between Underground Hoppers and Conveyors

Slide gates are widely used to regulate the discharge capacity by controlled opening and closing of the chutes outlets (Figure 203, Figure 204).

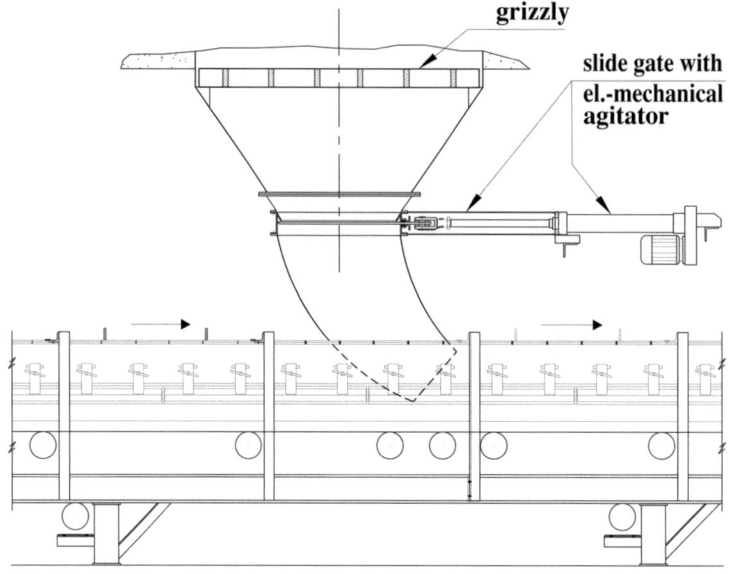

Figure 203. Underground hopper with slide gate, feeding discharge conveyor.

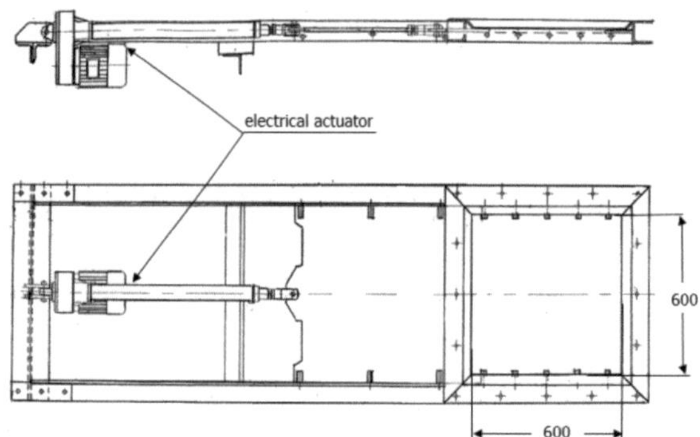

Figure 204. Slide gate controlled by electrical actuator.

A slide gate can be controlled open or closed by the pneumatic, hydraulic, or electrical actuator. The advantages and disadvantages of each option are considered below:

Pneumatic Actuators (Pneumatic Cylinders):

Advantages:

1. Compressed air is supplied from existing net that has an air pressure of 6.0 bars to 8.0 bars.
2. In case of failure of electricity supply, the solenoid valve keeps the horizontal slide gate in its last-used position.
3. Pull–push forces of a pneumatic cylinder cover most of the material handling requirements.
4. Additional positioning sensors (e.g. proximity switches) can provide intermediate positioning of the gate.
5. The cost of a pneumatic actuator is lower compared to other actuators.

Disadvantages:

1. In cases of compressed air supply failure, there is no power to move the horizontal blade (knife) to close or open the chute.
2. The pneumatic cylinder retracted in a vertical position cannot long hold the vertical load after compressor failure because of gradually reduction in air pressure.
3. Pressure losses, cleaning and drying of the compressed air, all this raises the Opex and Capex of the pneumatic systems.

Hydraulic Actuators (Hydraulic Cylinders):

Advantages:

1. Hydraulic actuators ensure large pull–push forces because of the high oil pressure provided by hydraulic pumps.
2. A hydraulic actuator can hold the pressure/force after hydraulic pump failure.
3. Hydraulic pumps can be located a considerable distance away from the actuator because losses in oil pressure are minimal.

Disadvantage:

The hydraulic actuator is only one part of the system consisting of oil pump, oil tank, filter, heat exchanger, oil pressure regulator, and so on. Installation of such a system results in a relatively high Opex and Capex.

Electrical Actuators

Electrical actuators differ from pneumatic and hydraulic actuators. An electrical actuator is driven by a gear-motor that rotates a screw bar which has continuous thread along its length. The non-rotating nut (or ball nut) rod moves linearly by the rotating lead screw. The electrical actuator is supplied with a few built-in internal limit switches.

Advantages:

1. The electrical actuator can be chosen to provide the *required push–pull forces* and the *required speed*.
2. Electrical actuators offer the highest-precision control of positioning using internal and outer limit switches.
3. In the case of electricity failure, the slide gate remains in its last-used position.
4. Modern electrical actuators are equipped with a manual bypass for opening and closing of the gate in the case of electricity failure.

Disadvantage:

The cost of an electrical actuator is slightly higher than the cost of a pneumatic actuator or hydraulic actuator.

3.3.3.3 Separation Walls

The advantage of rectangular storages in the ability to store simultaneously several types of bulk materials (or different grades of one bulk material) within the same storage. The different bulk materials can be stored without separation walls as partitioned piles with minimal distance between them (about 1.0 m), but such storing results in the misuse of a high percentage of the storing capacity. The bulk materials can also be stored in piles, separated by movable or fixed walls (Figure 205) to increase the store capacity (see the calculations below).

The movable wall must be heavy enough to prevent its slippage when the bulk material is loaded only on one side of the wall.

A typical movable separation concrete wall 6 m high, 1.25 m wide and 12.5 ton weight is shown in Figure 205 and Figure 206.

Figure 205. Typical 6 m high movable wall separating white potash from red potash.

The bulk material should be stored to a level below the top of the concrete-retained side walls and separation walls to prevent the material from spilling over.

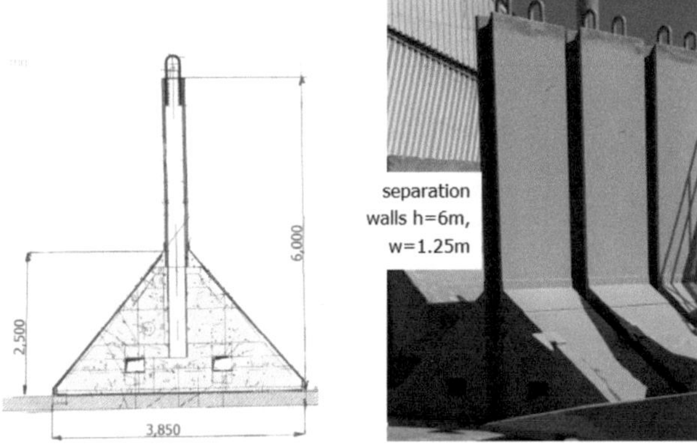

Figure 206. Example of separation wall (see also Figure 205).

The calculations below show how a separation wall (depending on its height) increases the total storing capacity of a storage when two different materials are stored separately. Storing of three or four different bulk materials in one storage makes the advantages of using separation walls even more significant.

The calculations were carried out for a typical rectangular Dome-type storage 80 m in length, 50 m in width, and with retaining walls 4.5 m high. The separation walls of different heights (h=0.0 m –without a separation wall at all, h=4.0 m, and h=6.0 m) were "installed" in the middle of the storage. The calculations were based on a 30°- repose angle of the bulk material.

Efficiency of the Separation Walls

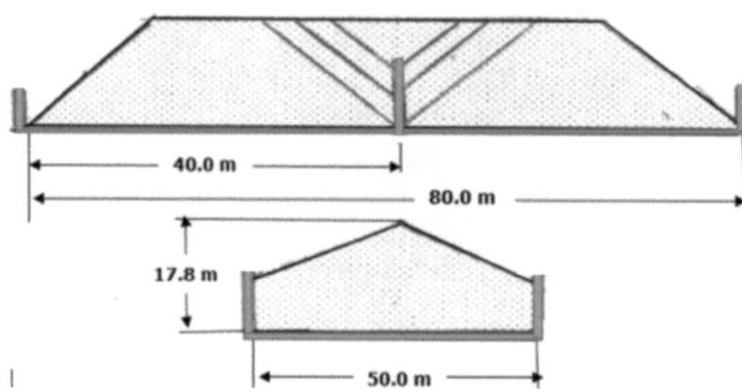

For one type/grade of stored bulk material, the storing capacity (V) is:

$V \approx 31{,}600\ m^3$ at the height of pile H = 17.8 m.

1. For two types/grades of material stored without a separation wall height of 0 m - it means two piles without separation distance between them:

$$V_1 \approx 6{,}130\ m^3 + 6{,}130\ m^3 = 12{,}260\ m^3.$$

2. For two types/grades of material stored with a separation wall height of 4 m:

$$V_2 \approx 9{,}300\ m^3 + 9{,}300\ m^3 = 18{,}600\ m^3.$$

3. For two types/grades of material stored with a separation wall height of 6 m:

$$V_3 \approx 11{,}000\ m^3 + 11{,}000\ m^3 = 22{,}000\ m^3.$$

Summary of the Calculations:

1. The storing capacity of *one* grade of bulk material (without an intermediate wall,) is 31,600 m^3 and this is the maximum store capacity, or 100%.
2. For the storing of *two* grades of material *without* an intermediate wall in the middle of the storage h=0 m and *without* a separation distance between two piles, the total store capacity of the storage is 12,260 m^3 or about 38% of the maximum capacity.
3. For the storing of *two* grades of material with an intermediate separation wall h=4.0 m, the total capacity is 18,600 m^3 or about 60% of the maximum capacity.
4. For the storing of *two* grades of material with an intermediate separation wall h=6.0 m, the total store capacity is 22,000 m^3 or 70% of the maximum capacity.

The separation walls not only significantly increase the store capacity of the storage but also prevent contamination one bulk material in the other.

3.3.3.4 Trippers and Tilting Probes.

Trippers

The overhead belt conveyor with a tripper is the most common solution for evenly loading of a rectangular bulk storage.

A cross section of a rectangular shed-type storage loaded by overhead conveyor with a tripper is shown in Figure 207.

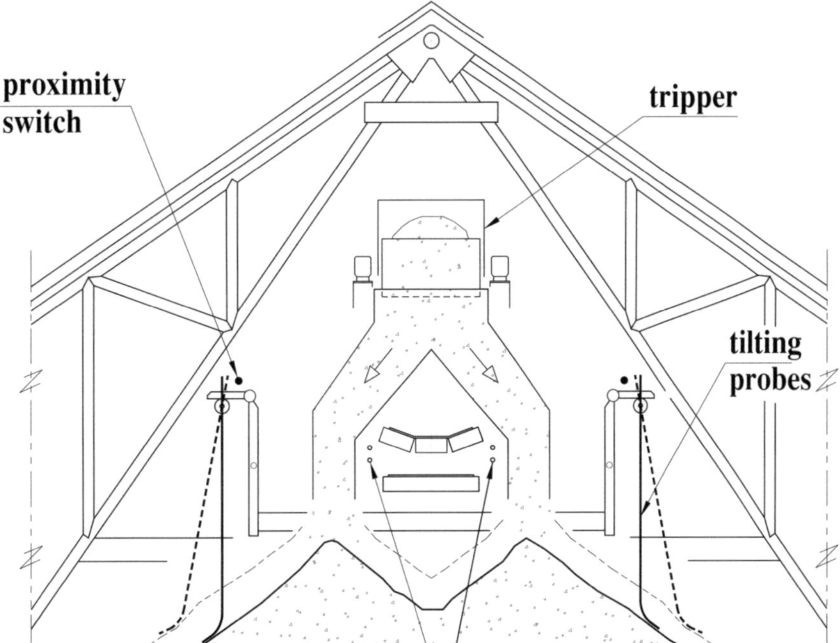

Figure 207. Cross section of overhead belt conveyor with tripper and tilting probes.

The tripper moves (or moved) above the overhead conveyor, on rails installed on the frame of the conveyor, and discharges the conveyor by elevating the loaded belt to 2 m to 3 m height, and then dropping the bulk material through a single chute or double chutes of the tripper into the storage.

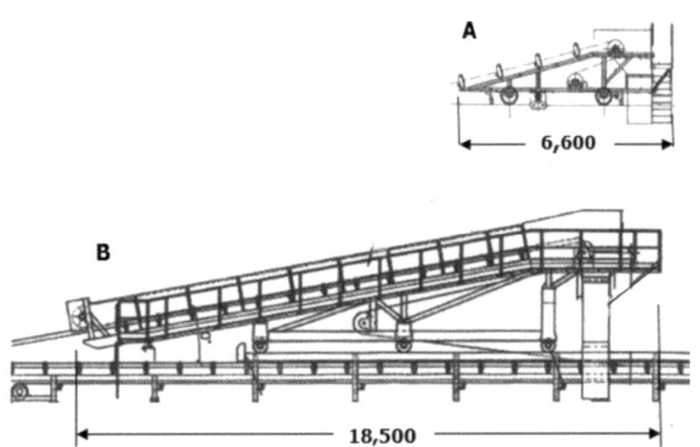

Figure 208. Two trippers designed by the two project companies for one B1200 belt conveyor (the trippers are shown at the same scale).

There are no International Standards or detailed recommendations for design of the trippers. The design of a tripper depends only on the experience of the project engineer.

The slope angle of the tripper according to our experience is 14°÷15° for the large majority of bulk materials as length of a tripper depends on this angle.

Fig. 208 shows two trippers (at the same scale!) designed by two project companies for two different terminals, for the same B1200 belt conveyor at the same capacity and with the same bulk material. Tripper A designed by Ring Projects (Israel), was manufactured and has been successfully operating for decades, whereas tripper B was recently designed by one of the European project companies. A comparison between the two options (dimensions, weight, maintenance and the cost) clearly indicates the advantages of tripper A.

The schemes of operating trippers are presented in Figure 209 and Figure 210.

Self-moving trippers are driven by drive units (a drive unit includes a 4.0 kW gear-motor) installed on the travel wheels of the tripper. The electric motors of the drive units require power to be supplied along the entire route of the moving tripper. The power can be supplied by a festoon system or by a motorized cable reel (Figure 210). These systems (especially in the storages contained dusty and/or sticky bulk materials) require regular inspection, cleaning, and maintenance.

The optimal alternative to the self-moving tripper is the tripper that is pulled by a wire rope that is actuated by a simple fixed friction winch (Figure 207). Such a system will be described below.

The principle of the tripper pulling system is based on Euler's law. A wire rope 16 mm in diameter wraps around the drive and idle pulleys three times to make a huge (3 × 190°) wrap angle and that helps to pull the tripper in two directions, as the rigging screw (Figure 211) is used to tense the rope (T_2) and to prevent the slippage of the rope on the drive (friction) pulley. The principle scheme of a tripper-pulling system driven by a friction winch is shown in Fig. 211. This drive system designed to pull existing tripper (the belt conveyor with a capacity of 1,250 tph) along a storage 130 m in length (Figure 212).

The tripper travels above the overhead conveyor at a speed of 0.3 m/s and is stopped by magnetic sensors that are assembled on intermediate discharge stations every 3.0 m. Each station is equipped with two mechanical probes (e.g., a tilting probe activated proximate switch, Fig. 210) to control and limit the maximum permissible height of the pile. After the pile has reached the maximum height, the tilted probe actuated proximity switch and the program starts to move the tripper automatically to the next empty discharge station (Figure 206).

If all stations are filled up, the tripper "informs" the control room and starts to move from station to station for five minutes only. After this, it stops the overhead conveyor. The range of the loading stations for each grade of bulk material should be programmed by the operator from the control room at the beginning of the storage loading operation.

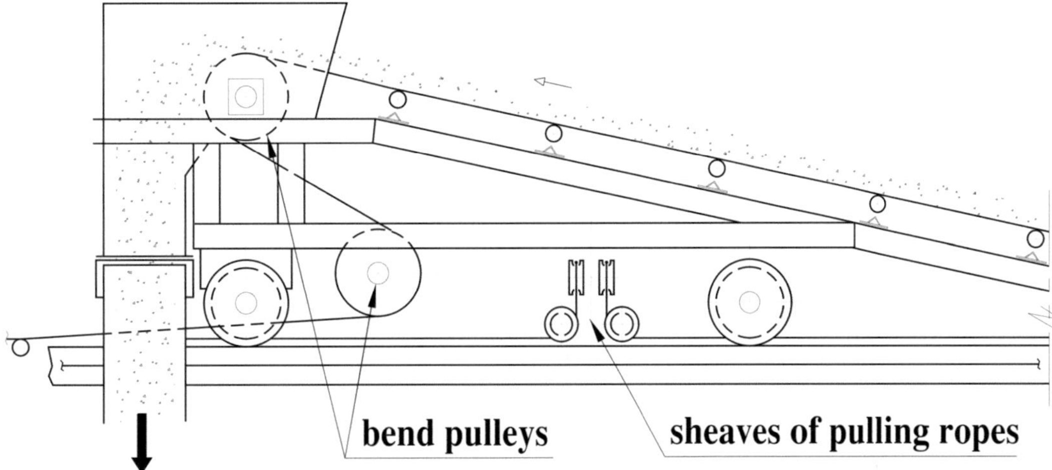

Figure 209. Tripper pulled by steel rope via friction winch (see also Figure 211, Figure 212).

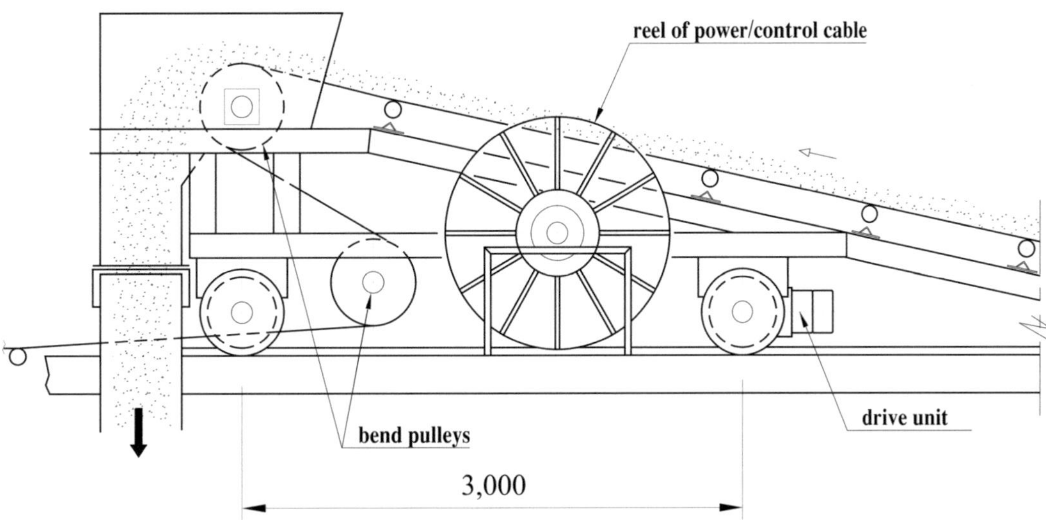

Figure 210. Motorized tripper equipped with drive units installed on travel wheels, and the motorized reel for the cables of control/power supply.

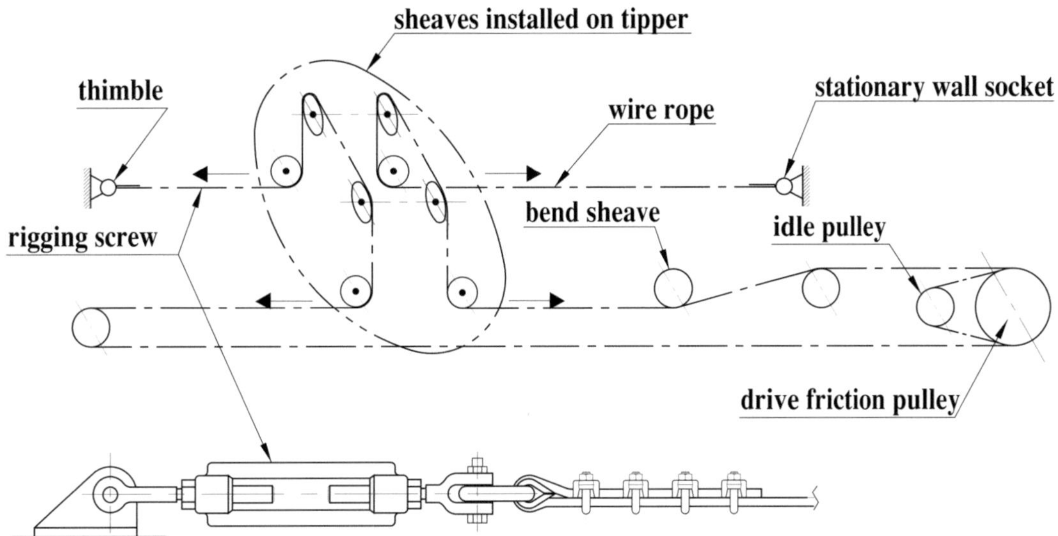

Figure 211. Scheme the friction winch system for pulling a tripper.

The rigging screw (Figure 211) is used to tense the steel rope (T_2) to prevent the slippage of the rope on the drive (friction) pulley.

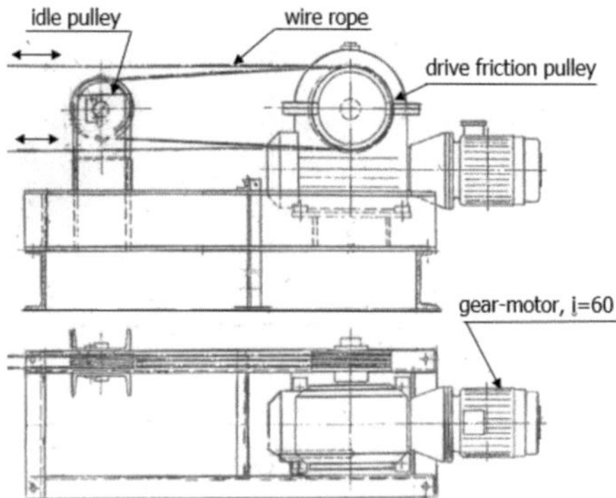

Figure 212. The 4.0 kW fixed friction winch pulling a B1200 tripper (1,250 tph capacity) along a rectangular storage 130 m in length at a speed of 0.3 m/s.

How to use two trippers installed one after another on the same belt conveyor?

The operation of an overhead belt conveyor with a single tripper, loading multi-bay storage, stored different bulk materials, often there is a problem: the tripper, moving over bays, continuously pours out the bulk material, contaminating other bays. This problem arose when the Tsefa storage (four bays plus one bay/ truck-discharge passage in the middle) was designed and built with one tripper. The overhead conveyor is fed directly from the overland belt conveyor (18.2 km long Sdom–Tsefa Cable Belt conveyor), so the stop of the overhead conveyor (to move the tripper from bay to bay without contamination of intermediate bays) causes undesirable stops not only the overland conveyor, but all the chain of conveyors feeding the overland conveyor (Figure 211).

The solution to the problem

A second tripper (no. 2) was installed inline with tripper no. 1 (Figure 212). Tripper no. 2 was equipped with a gravity diverter. The diverter allows to load the bay, or transfer the material back to the belt along the direction to the second tripper no. 1 that is already installed above the bay that should be loaded.

Now, for example, to load bay no. 5 (after bay no. 1 was filled with the same material), the second tripper is automatically moved to bay no. 5, as the first tripper stops the loading of bay no. 1 and, and using the diverter, transfers the material forward on the belt towards tipper no. 2, that begins to load bay no. 5 without contamination of bays no. 2, no. 3, and no. 4.

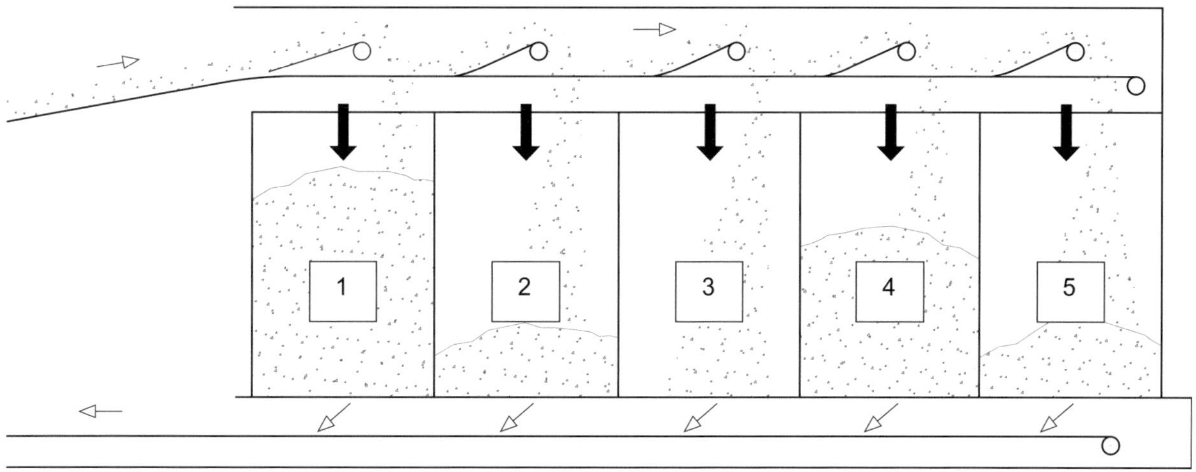

Figure 211. The tripper no. 1 moves from bay no. 1 to bay no. 5, contaminating bays no. 2, no. 3, and no. 4.

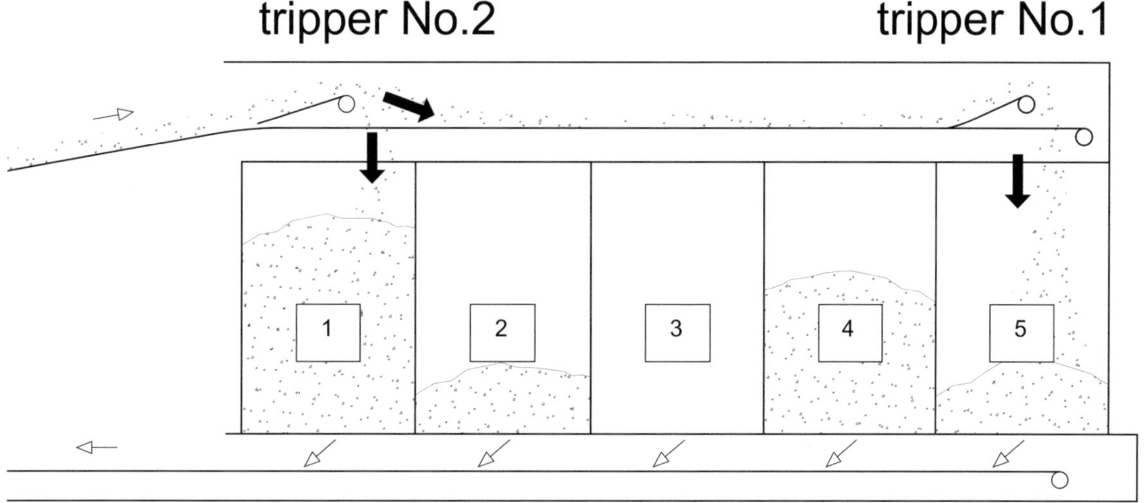

Figure 212. Two trippers installed inline on the same overhead belt conveyor.

Figure 213. Two trippers installed in line: tripper no. 2 (A) and tripper no. 1 (B) (see also Figure 211 and Figure 212).

All trippers, operated in this facility, are pulled by rope friction winches, and the festoon is used only to supply compressed air/control cable to the pneumatic cylinder, installed on tripper no. 2 and actuating the gravity diverter (Figure 213, A).

Advantages of the friction winch over the motorized tripper:

1. The rope friction winch is a fixed compact drive unit, located in a convenient place for inspection and maintenance.
2. The tripper is drawn by pulling rope from both sides of a tripper evenly (see yellow arrows in Figure 211), with no derailing problems.
3. A power supply to the moving tripper is not required.
4. We found that the friction winch is the optimal solution: simple, reliable, low-maintenance and inexpensive.

Disadvantages:

No disadvantages were discovered for long and successful operating time in storages full of heavy dust. All our old trippers with motorized wheels were reequipped to be pulled by rope friction winches.

Tilting Probes

The automatic even loading of a storage by overhead conveyor with a tripper requires the installation of sensors/probes (mechanical or electronic) to control the maximum permitted height of a pile in order to prevent overload and damage to the overhead gallery (Figure 207).

We had tried various standard sensors (lasers, rotary paddles, vibratory level-control measuring devices, and so on) that successfully operate in storages containing non-sticky powders and granular solids, but all of them were not reliable enough to operate in our storages filled of fine, sticky dust. Using the trial and error process, we developed, manufactured and installed in all our storages the simple mechanical probes shown in Figure 214 and Fig. 215.

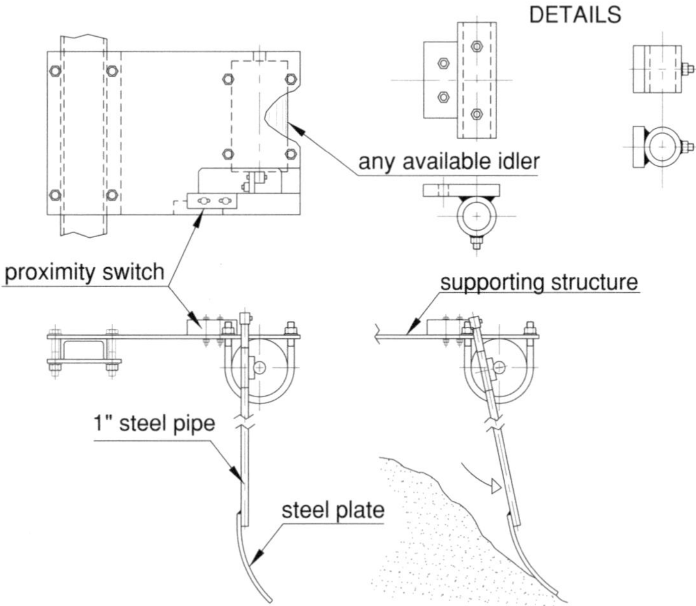

Figure 214. Construction of the tilting probe.

Each tilting probe consists of one standard conveyor idler, 1" steel pipe, and one proximity switch. The idlers are fixed to the handrail from both sides of overhead gallery at the discharge stations. The pipe is attached to the axis of the idler and could be freely tilted. During the loading of the storage, the pivoted pipe equipped with lower curved steel plate, is in its upright position. When the pile approaches its maximum height, the material begins to tilt lower section of the pivoted pipe outward as upper section of the pipe, tilted in the opposite direction, actuates a fixed proximity switch. The switch transmits the signal to the operational program, and this begins to move the tripper automatically to the next empty discharge station.

Figure 215 shows the suspended tilting probes (shown in Figure 214) installed on the discharge stations of overhead gallery to ensure a uniform distribution of a bulk material along the length of the storage. The probes have been operating since 1988 with no maintenance and with no replacement.

Figure 215. The tilting probes (see also Figure 214) installed on the discharge stations of the overhead conveyor of potash storage.

Cleveland Cascade Chute

To reduce heavy dusting during loading of a storage by reducing the material free-fall velocity, Cleveland Cascade telescopic chute can be hung from overhead conveyor gallery at the first discharge station (the suspension of the tripper chute/s is the dangerous, unsafe solution). So, the first heap can be piled up, practically, dustless, but continuation of the loading from the first pile onwards (already without the chute) will generate a lot of dusting.

Additional disadvantages of using Cleveland Cascade chutes in storages:

1. The extended telescopic chute in a dusty environment can be caught and pulled by the shovel. The pulling of the chute will cause damage to the overhead gallery and can be dangerous for safety reasons.

2. The occasionally blockage of the Cleveland Cascade chute (the total weight of a fully extended and blocked chute can reach 20 t) will cause local overload of the structure of the overhead gallery.

3. Two Cleveland Cascades are required for usual two-sides discharging tripper (Fig. 207).

4. High Opex and high Capex.

3.3.4 Rectangular Storages Loaded by Overhead Conveyor with Tripper and Unloaded via Floor-level Side-located Belt Conveyor

The typical covered storage, discharged by a side-located belt conveyor installed on the free from material area of the storage, on the floor level, is shown in Figure 216.

The horizontal belt conveyor is located along the entire length of the storage and transfers the material onto the next takeaway belt conveyor. The shovels load the conveyor through special hoppers.

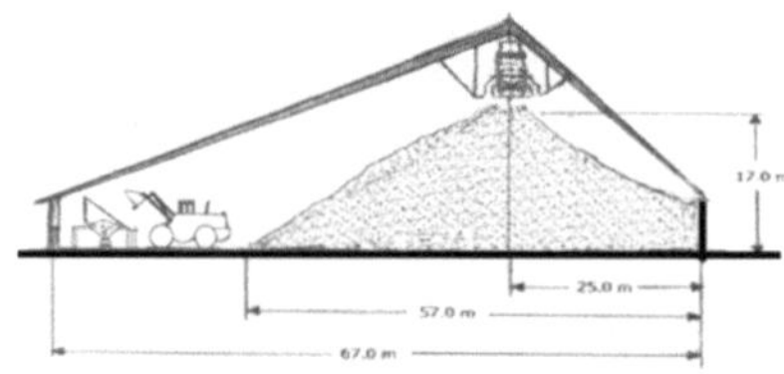

Figure 216. Cross section of typical rectangular storage loaded by overhead conveyor with tripper and discharged through side-located floor-level belt conveyor.

The hoppers are installed above the discharge belt conveyors, and the rate of the conveyor loading is controlled by the cross section of the hopper outlet (Figure 217, Figure 218) or using of belt feeder installed on the bottom section of the hopper (Figure 220).

The discharge hopper can be installed at various locations of the storage, as is shown in Figure 217.

Figure 217. Discharge hopper with belt feeder installed inside the storage.

Figure 218. Discharge hopper installed above the side-located belt conveyor.

Similar storages are in use in Finland (port of Kotka) (Figure 219, Figure 219.1).

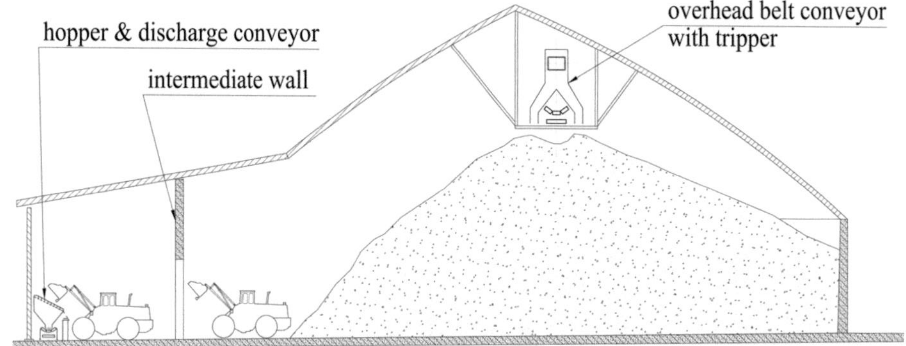

Figure 219. Cross section of a covered storage for bulk urine (Finland).

Figure 219.1. The intermediate wall of the urine storage shown in Figure 219.

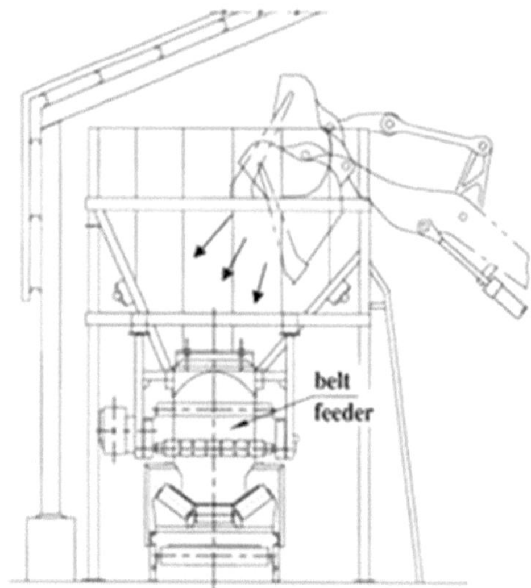

Figure 220. Design of a discharge hopper with a belt feeder.

A wide, free from bulk material storage area makes it possible to use of trucks, loaded by shovels, for unloading a storage (Figure 221).

Figure 221. The storage (see Fig. 216) discharged by trucks loaded by shovels.

The storages, shown in Figure 222, were recently designed for one of the European ports. Storage A is for non-free-flowing bulk materials discharged by shovel/s into the side-located hoppers to floor-level belt conveyor, and storage B is for free-flowing bulk materials (30%÷50% to be discharged by gravity, and the remaining material to be discharged by shovels into the underground hoppers feeding underground belt conveyor).

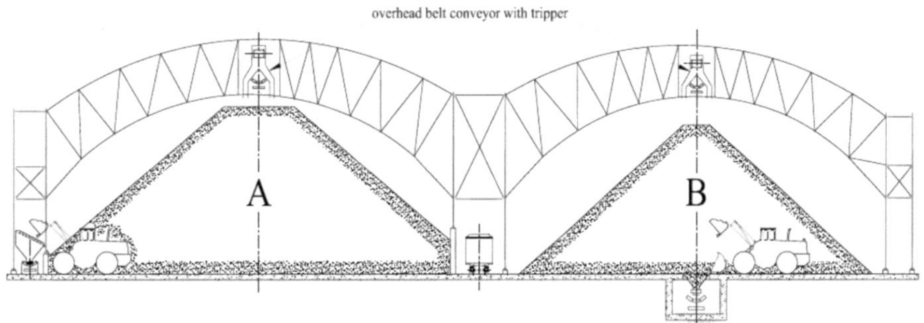

Figure 222. Cross section of two rectangular storages.

3.3.5 Rectangular Storages Loaded by Overhead Conveyor with Tripper and Unloaded via Side-located Underground Belt Conveyor

As an option, a discharge conveyor can be installed in the area free from material, within a side-located underground gallery (Figure 223).

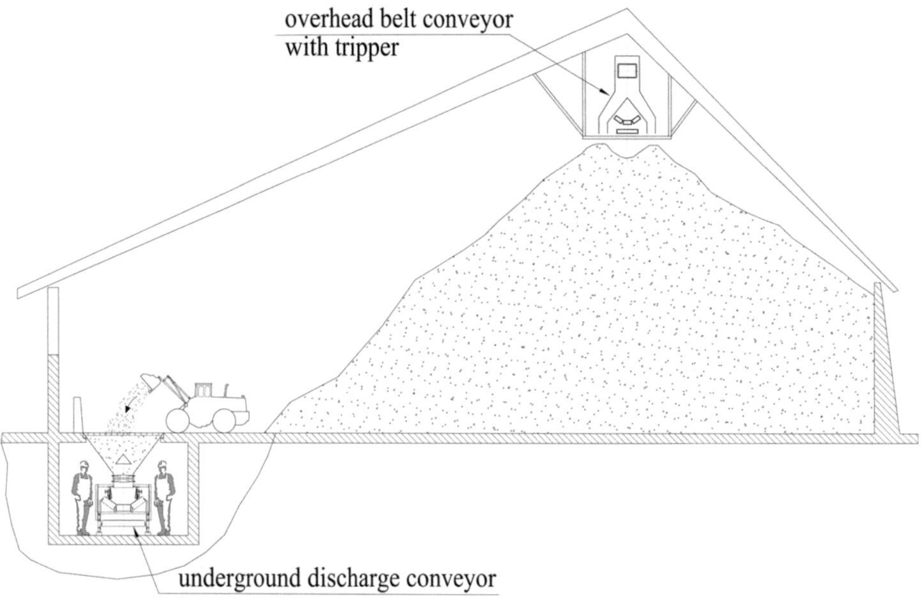

Figure 223. Storage is unloaded by shovels through side-located underground discharge belt conveyor.

The storages with side-located underground conveyors have the higher discharge capacity per shovel (by 15%÷20%) compared to storages with side-located floor-level conveyors. The reason: it takes less time for a shovel to drag bulk material and then drop it into an underground hopper than it takes to fill a bucket with the material, raise the bucket, drive to a hopper, and pour out the material into the hopper from above.

Figure 224 shows three storages with side-located underground discharge conveyors for bulk fertilizers that were designed for a port facility in Thailand.

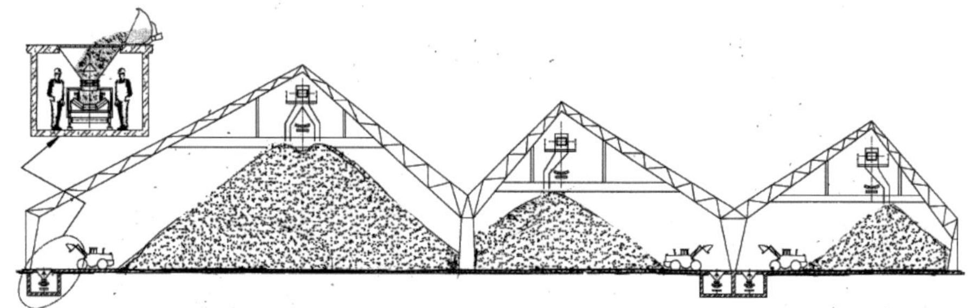

Figure 224. Storages with side-located underground discharge conveyors.

Advantages of storages with side-located floor-level conveyors:

1. Easy to maintain.
2. No problem with discharge of non-free-flowing material. One Caterpillar 988 shovel provides up to 650 tph average discharge capacity. For higher capacities, several shovels and hoppers are required.
3. Different bulk materials (separated by walls) can be stored in one storage.
4. No problem in cleaning spillages from takeaway conveyor (this problem does appear when takeaway conveyor installed in side-located underground gallery).
5. Relatively low Capex.

Disadvantages of storages with side-located floor-level conveyors:

1. The footprint/cross section of these storages is larger by 15% than that of a Dome-type rectangular storage with central underground conveyor having the same bulk material cross section.
2. All stored material must be discharged by shovels.
3. To reach the minimum discharge capacity required today for shiploading (1,000 tph to 1,500 tph, and more), two or three shovels must operate concurrently, resulting in significant Opex.
4. Side-discharged storages usually have one (discharge) side of the storage open, resulting in significant dust emission. The new international and local environmental rules require that storages for dusty bulk materials to be enclosed.

3.3.6 Rectangular Storages Loaded by Overhead Conveyor with Tripper and Unloaded by Portal Reclaimer

A cross section of the storage loaded by an overhead belt conveyor with a tripper and unloaded by a travelling portal reclaimer is shown in Figure 225.

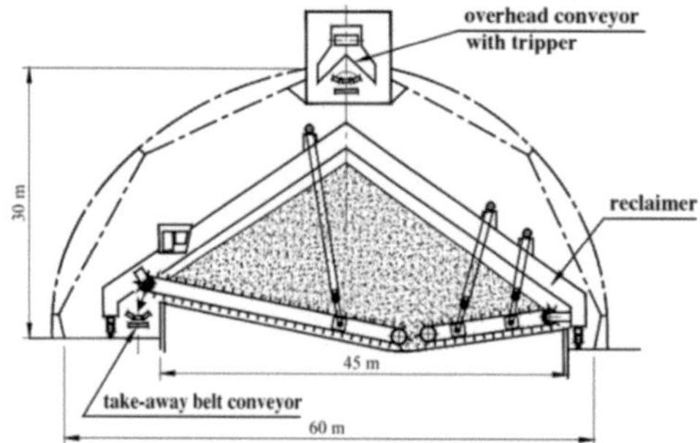

Figure 225. Typical storage with overhead conveyor with tripper and portal scraper reclaimer.

The scraper-type portal reclaimer is an endless scraper chain equipped with blades or buckets rotating continuously around a scraper boom and supported by an A-frame portal travelling along the storage on rails.

Figure 226. Storage with overhead conveyor and reclaimer (TAKRAF Tenova).

The A-frame of a reclaimer usually carries two scraper booms, forming one discharge unit.

The discharged bulk material is scraped by the blades from the pile to the side, where it is then poured out on the discharge belt conveyor (Figure 225, Figure 226).

Scraper-type portal reclaimers are fully automated, remote-controlled machines.

3.3.7 Rectangular Storage Loaded by Travelling Stacker and Unloaded by Portal Reclaimer

A cross section of a rectangular storage with a travelling stacker and reclaimer is shown in Figure 227.

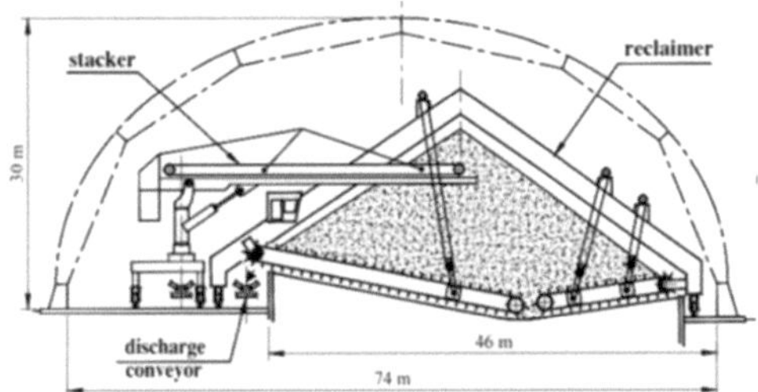

Figure 227. Cross section of a typical storage with stacker and scraper reclaimer.

Figure 228. Example of loading of storage by stacker (Bedeschi, Italy).

Advantages of storages unloaded with scraper reclaimers:

1. Automatic discharge of the storage.
2. Uniform and controlled rate of the discharge.
3. No shovels are required to discharge or to clean a normally operating storage.

Disadvantages of storages unloaded with scraper reclaimers:

1. The footprint of one such storage is 20% to 60% larger than the footprint of a rectangular storage with a central underground conveyor having the same cross section of the stored bulk material.
2. Complex mechanical systems are located inside the storage. Dusty, sticky, corrosive, and abrasive bulk materials decrease the reliability of the travelling, mechanical and control systems.

3. Long breaks between discharge operations (e.g., the waiting for vessels coming to be loaded in the port) may lead to mechanical failures when the equipment must be re-started, especially when the bulk material is dusty and sticky.

4. Maintenance of a scraper-type portal reclaimer must be carried out by the skilled mechanical, electrical, control, and hydraulic teams, resulting in high Opex.

5. In case of a breaks of reclaimer or one of takeaway conveyor, the storage cannot be discharged by any other means.

6. The cost is high and higher than the cost of other types of storage discharge systems.

The use of a reclaimer discharge bulk storage can be justified in the case of handling dangerous or dusty bulk materials where the use of shovels may be risky and unsafe for drivers, or where the comparison between Opex and Capex shows clear advantage of using reclaimer.

The stackers operated in the storages are large travelling rail-mounted portal machines used to pile bulk materials.

In principle, the stacker is a large "motorized tripper", with a head pulley installed up to 25 m high. The bulk material is transferred by a stacker to the tail section of the boom belt conveyor. The stacker can swivel and tilt the boom conveyor allowing it to reduce the material free fall height during pouring out bulk material into any point in the storage with reduced dust emission (Figure 228).

Advantages of a stacker:

1. Mobility and possibility to load bulk material on any spot of the storage.

2. The stacker can tilt the boom conveyor downward in order to lower the height of the material drop and reduce dust generation during loading of the storage.

3. Fully automatic operation.

Disadvantages of a stacker:

1. The regular inspection and maintenance of such machine must be carried out by a professional team of mechanical, electrical, control, and hydraulic specialists.

2. High Capex: the cost of a stacker is usually high and can be equal to and even higher than the cost of a reclaimer.

Example from the site:

In 1979 a large rectangular storage for fine phosphate was built in port of Ashdod. The storage is loaded by a stacker and unloaded by a portal reclaimer (Figure 227).

Ten years later, a second phosphate storage was built in this port. Considering the experience of the previous storage, the new storage is unloaded by reclaimer, as the first one, but it is loaded by an overhead conveyor with a tripper instead of a stacker (Figure 225). The new storage loading system (that is less-expensive than the previous one) significantly reduces maintenance expenses.

3.3.8 Small Storages Unloaded via Underground Conveyors

Such small storages with underground belt conveyors are normally built to ensure high discharge capacities and to reduce attrition of fragile bulk materials as a high percentage of the stored material is discharged by gravity.

For example, the 10,000 tons multi-bay storage built in 1986–1987 for the Tsefa wagon-loading facility (Israel). The storage consists of four separate bays, each 20 m in length and 2400 tons capacity, loaded with one overhead conveyor (up to 800 tph capacity) with two trippers in line. The storage is unloaded by one underground discharge conveyor of 1,200 tph capacity (Figure 229, Figure 230).

The high discharge capacity is needed because the potash is transported directly to the loading tower where the wagons are loaded from 80-tons silo with average capacity of about 1000 tph. The loading one 65-ton payload wagon takes 4 min to 4.5 min.

The 5-m wide truck passage, located in the middle of the storage, with one underground hopper, can be used as a discharge pit for tipper trucks or can store about 700 tons of bulk material (Figure 230).

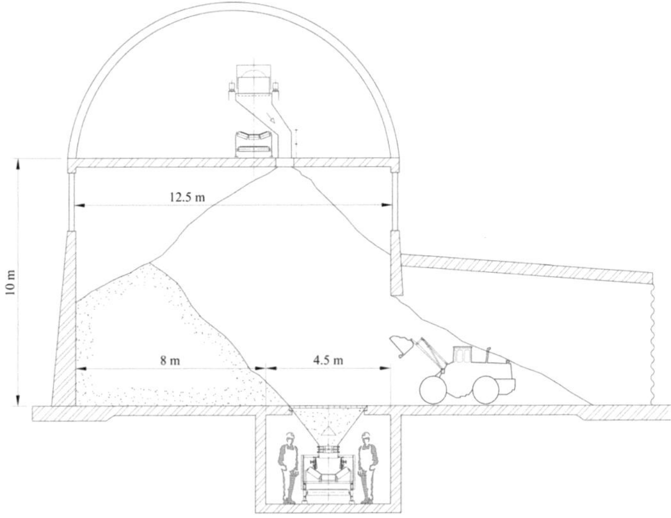

Figure 229. Cross section of Tsefa multi-bay storage.

Figure 229 shows that the underground conveyor was built asymmetrical to the centre line of the storage. This was done in order to facilitate approach of the shovel to underground discharge hoppers. The underground hoppers in this storage were installed in line, side by side, so most of the potash is discharged onto underground conveyor by gravity.

All rectangular storages have side gates which should be wide and high enough to allow a shovel to enter and exit.

The high-speed roll-up doors were chosen for the Tsefa storage gates (and for the bulk storages at the ports of Ashdod and Eilat) because they have proved to be reliable and low-maintenance equipment.

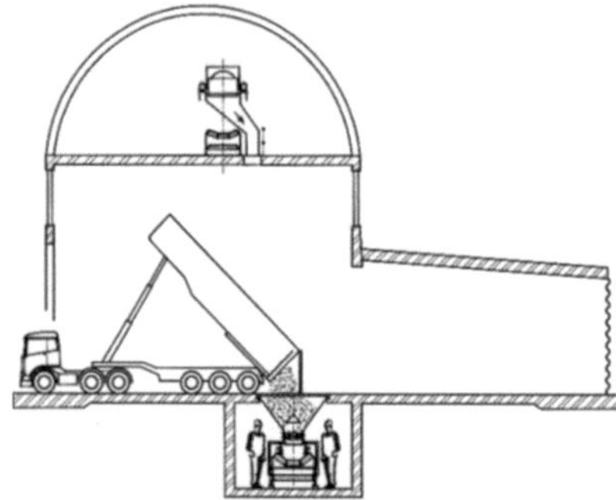

Figure 230. The four-bay potash storage with an additional bay (which in use as a truck- discharge passage) in the middle of the storage (bay no. 3).

3.4 Storage Bins

Big steel storage bins (1,000 t ÷ 2,000 t) are used for storing free-flowing bulk material (grain, coal) or easily fluidized powders (cement, fine phosphate) (Figure 231).

The bins are manufactured in various configurations.

For easily agglomerated bulk materials as potash and salt, most bins have a storing capacity up to 200 tons.

This limitation is due to the danger of clogging, or bridging of bulk materials in the bins (especially after relatively long breaks between discharges) resulting in stop the bulk material flow. Cleaning such a bin is a very hard and time-consuming operation, and the bigger the bin, the bigger the problem.

Bins used for wagon and truck loading are normally equipped with bin activators (to prevent ratholing and bridging) and with dedusting telescopic loading chutes (e.g., Cleveland Cascade or similar).

Figure 231. The industrial storage bins for the storing free-flowing powders.

Free-flowing bulk materials (such as grains) do not usually cause the discharge problems. A small electrical vibrator, installed on the lower cone at a distance 1/3 away from the outlet, or a couple of fluidizing pads for powders, help re-start the mass flow even after relatively long breaks.

The dimensions of a bin outlet opening for loose, free-flowing materials can be calculated as:

$$A = M /\{\varrho\, [B \times g /(2(1 + m) \tan \theta)]0.5\}$$

where A is the outlet area (m^2); ϱ is the bulk density (kg/m^3); M is the mass flow rate (kg/s); B is the diameter of circular or width of slotted outlet (m); $m = 1$ (for circular opening) and $m = 0$ (for slot-shaped opening); g is the gravitational constant (m/s^2); and θ is the channel angle (gradient).

The problem is how to choose parameters of the bin that will make possible to reach required discharge capacity of the *non-free-flowing* bulk material.

Bridging (Figure 232, A) and rathole (Figure 232, B) are the silo discharge problems with which the engineer meets very often.

The flowing characteristics of bulk materials were considered in Section 1.

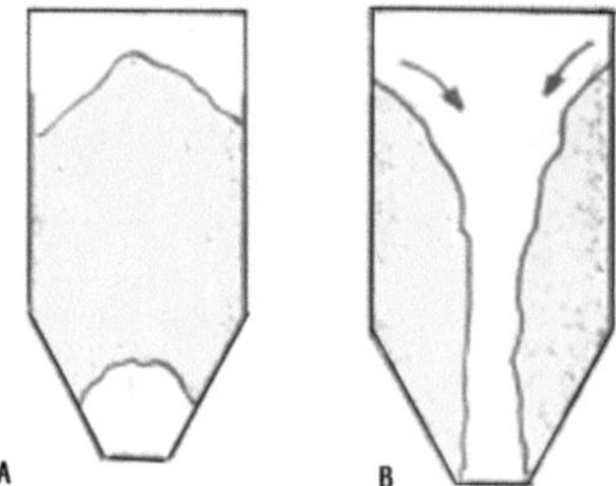

Figure 232. Bridging (A) and ratholing (B) discharge problems.

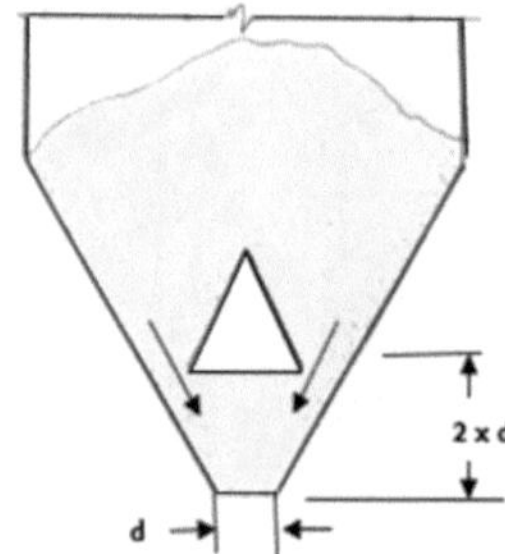

Figure 233. The conical insert (especially vibrated) helps prevent bridging.

A simple and effective device that improves a bin discharge capability is a bin activator (Figure 234, Figure 235). Bin activator is the simple device of conical shape suspended to the bottom section of the silo through flexible connection. The cone vibration activates the bulk material on the outlet of the silo.

Thus, the bin activator not only increases the critical diameter of the silo outlet (D_c, Figure 2.2) but at the same time uses vibratory action to activate mass flow.

In our practice, to ensure stable discharge of highly hydrophobic and very sticky bulk material from silo 6.0 m in diameter and capacity of 150 tons, we recommended installing of 4.0 m in diameter bin activator and, in addition, putting in 5.9 m in diameter cylindrical flexible rubber insert (to avoid contact between the bulk material and the steel walls of the silo). The silo has been operating successfully since 1984 (Figure 236).

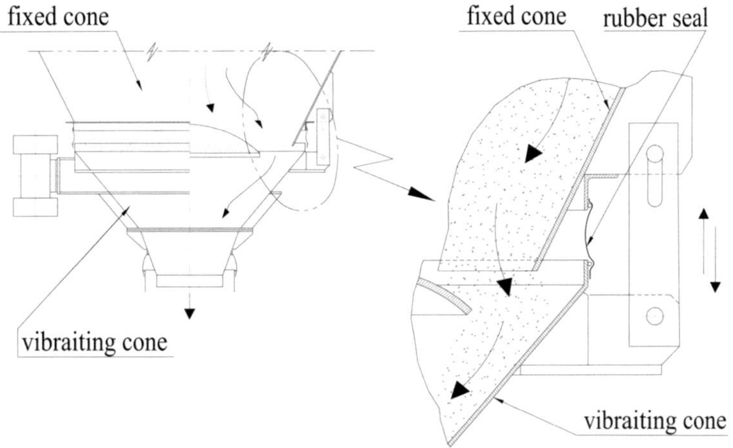

Figure 234. Construction of bin activator (IFE, Austria).

Figure 235. Bin activator (IFE, Austria).

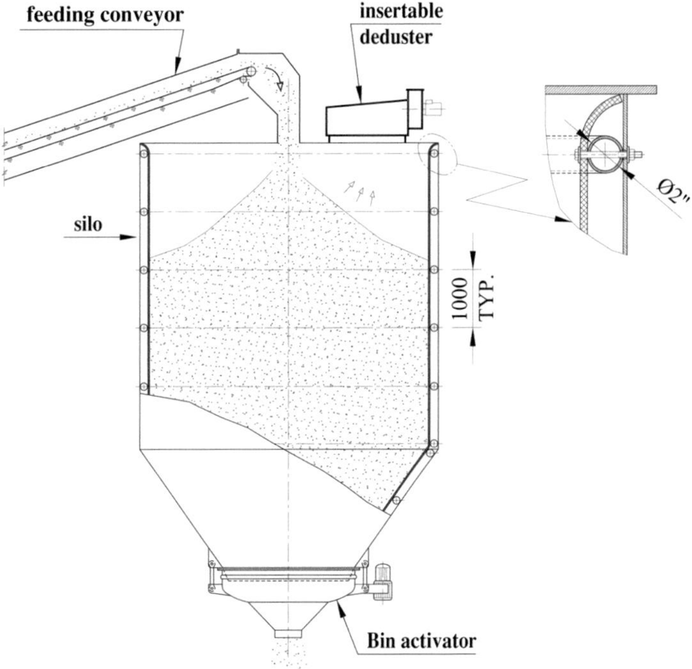

Figure 236. The use of a bin activator and internal rubber cylinder to stimulate the discharge of the silo.

CHAPTER 4.

Loading and Discharge of Trucks and Wagons.

General

Transportation of bulk materials (coal, grain, ore, fertilizer, etc.) by railway wagons and trucks is becoming more intensive and demanding each year.

All wagons/trucks loading facilities are designed and built as tailor-made facilities, for loading and unloading specific bulk materials, for specific types of railway wagons or trucks, for required capacities and for environmental requirements.

4.1 Loading of Trucks

Loading by shovels is the most common method of loading open trucks with non-dusty bulk materials. The shovel digs the bulk material from a storage pile or from open pile and poured it out straight into a truck (Figure 237). The shovel can also drop the material into an intermediate hopper equipped with a belt feeder loading the trucks.

Figure 237. Loading of bulk salt into an open truck by shovel.

Advantages of loading trucks by shovel:

1. Simple, low-cost operation.
2. No special equipment is needed.
3. Relatively high loading capacity (e.g., 200 tph to 300 tph for Caterpillar 950H shovels, 400 tph to 600 tph for Caterpillar 988H).
4. Double-trucks can be loaded without separation their bodies/boxes.

Disadvantages of loading trucks by shovel:

1. Heavy dust emission if the loaded bulk material is the dusty one.
2. Difficulties in distributing bulk material uniformly along the truck body.
3. Serious difficulties in loading the truck with the maximum permissible payload without weighing the truck, as the overloaded truck must be sent back in order to unload the surplus bulk material.
4. Installing of a truck on a fixed scale at an accuracy of ± 1%, requires additional investment and reduces operational flexibility of the loading operations. As an option, the hydraulic scale (e.g. VEI Group, Italy) can be installed on a shovel to weigh the material within the bucket.
5. Significant spillage during loading operations.

4.1.1 Truck-Loading Facilities

The large freight turnover requires the use of special facilities when they provide high loading capacities and prevent dust emission.

Figure 238 shows a truck-loading facility where the upper surcharge hopper fills the lower loading silo installed on load cells.

The feeding belt conveyor fills the surcharge hopper while the loading silo (as its upper slide gate is closed) fills the truck using a tilting chute (the tilting movement helps to distribute the material evenly along the truck).

When the truck is fully loaded, the lower slide gate of the loading silo is closed as its upper slide gate opens. The fully loaded truck drives away, and an empty truck takes its place under the loading silo. The material from the surcharge hopper continues filling the loading silo as weight of the material, getting into the silo, is controlled by load cells.

After the load cells signal that the required batch of material has been loaded, the upper slide gate of the silo is closed and its lower gate opens to begin the truck-loading operation. Further, the truck-loading cycle is repeated.

This operation requires close coordination between the operator and the truck driver because, to get uniform filling, the big truck must be shifted during the loading.

Figure 239 shows a truck-loading silo equipped with three telescopic Cleveland Cascade chutes. The feeding belt conveyor fills the silo installed on the load cells. The filling of the silo is re-started each time when the weight of material within the silo dropped to 40% of full store capacity of the silo: a silo that filled with only partial (< 40%) causes lack of uniformity in the loading capacities of the telescopic chutes (the central chute has a higher capacity than two side chutes), resulting in uneven loading of the bulk material along the truck body.

The Cleveland Cascade telescopic chutes installed on the outlets of a loading silo provide practically dustless filling of open trucks (Figure 239).

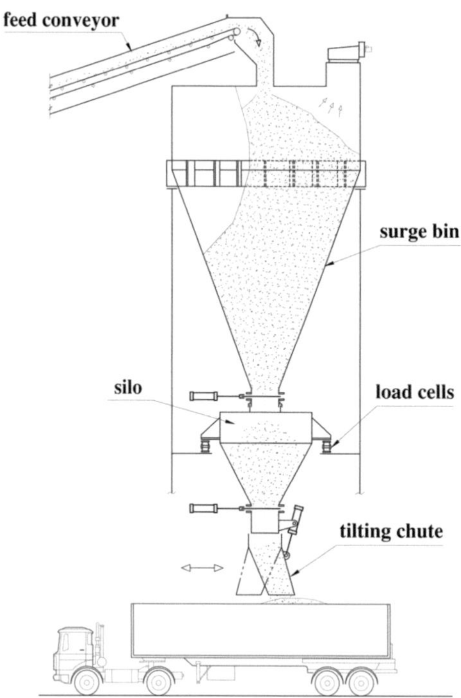

Figure 238. Loading of open trucks with pre-weighed batch of material, using tilting loading chute.

Since the loading of open trucks/wagons is accompanied by significant dust emission, most modern loading stations are enclosed and equipped with dedusting systems. Grain terminals are often built as special multi-bay truck-loading facilities for concurrent loading of six or more trucks through weighed loading silos that fed by overhead surcharge hoppers. The surcharge hoppers are filled by upper shuttle-type belt conveyor.

Figure 240 shows the loading of granular potash into a 20-ton open truck through two loading non-telescopic chutes with a total loading capacity of 650 tph to 700 tph. The potash is poured out (without a dedusting system) through the 80-ton loading silo, from a height of about 2 m.

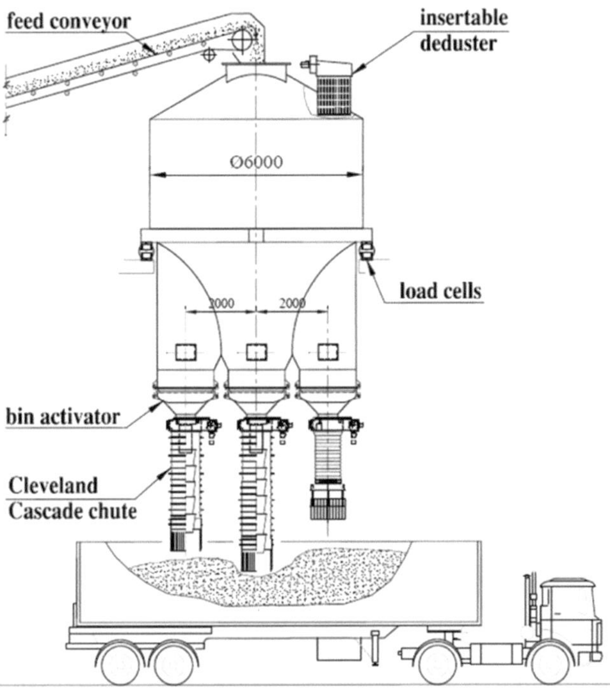

Figure 239. Truck-loading facility with three Cleveland Cascade telescopic chutes.

Figure 240. Truck loaded with granular potash without dedusting system.

Trucks can be loaded also by grab cranes during the ship-unloading. The grab crane unloads a vessel and fills movable truck-loading hopper (Figure 241).

Figure 241. Trucks loaded during ship unloading (Verstegen Grabs, Netherland).

4.2 Unloading of Trucks

Selecting the type of truck-unloading facility depends on the type of trucks and required unloading capacity. There are three main types of trucks for the transportation of bulk materials:

- Tipper trucks (self-unloaded), singles or double units.
- Non-tipper trucks with hinged back door (unloaded on tilting platforms).
- Tank trucks (or tanker trucks).

4.2.1 Unloading of Tipper Trucks

The tipper truck (Figure 242) is equipped with a hinged dump box that is tilted by a hydraulic cylinder to the angle at which the bulk material begins to slide down and porous through the opened, hinged back door.

The tipper truck can operate as a single-unit truck carrying payload of limited capacity (up to 32.5 tons or even more), or as a double-unit tipper truck.

Figure 242. Tipper truck unloads dusty ore (Spain).

The hydraulic cylinder, installed on a tipper truck platform to tilt the box, is operated by the truck driver actuating a fixed panel with joysticks (Figure 242).

The unload a single-unit tipper truck is relatively fast (1.5 min to 2.5 min), whereas the discharge of a double-unit usually takes between 4 minutes and 6 minutes. The volume of receiving hopper and the capacity of the feeder should be chosen in order to empty (or almost empty) the receiving hopper until the next truck will begin discharging.

When the pace of arrival of trucks relatively low, the discharge of *non-dusty* materials from tipper trucks can be carried out through small open discharge pits. The discharge capacity of such pit depends on the capacity of the feeder/takeaway conveyor.

The discharge of a dusty bulk material into an open pit is not recommended due to high dust emission.

To prevent dust emission during unload of trucks, the Samson enclosed movable truck dischargers, that widely used today, equipped with insertable dedusting systems.

Advantages of a Samson truck discharger (Figure 289):

1. Simplicity and reliability of operation.
2. High mobility as the discharger, with attached swivel belt conveyor, can be easily towed from place to place. A much more expensive self-propelled discharger can be ordered if it is necessary.
3. Built-in dedusting system is available.

Disadvantages of a Samson truck discharger:

1. Each double truck must be split and each part unloaded separately.
2. The manoeuvres of double tipper trucks while splitting, moving in and out of the Samson discharger and reconnecting the parts again, require significant space (free from other equipment!) on the wharf.
3. Low average discharge capacities of 200 tph to 400 tph.
4. High Opex and relatively high Capex.

4.2.1.1 Discharge Pits

The large fright turnover carried out by trucks, requires special, high-capacity, enclosed and dedusted (in the case of dusty materials) discharge pits. In port of Eilat, the tipper trucks are unloaded into enclosed pits with underground hoppers of about 30 m^3 in store capacity, equipped with low-speed belt feeders transferring the material to takeaway belt conveyors.

Figure 243 and Figure 244 show a discharge pit for bulk potash with the discharge capacity of about 750 tph.

The baghouse filter, at a capacity of 50,000 Nm^3/hr and with a dust emission of about 5.0 mg/m^3, was installed close to the discharge pit.

The screw conveyor, installed in the bottom of the baghouse hopper, transfers the collected dust back on to the takeaway conveyor or, optionally, fills big bags.

The enclosed discharge pit for fine phosphate ($d_{50} \leq 60$ μm) is dedusted by a baghouse filter of 105,000 m^3/hr capacity, and the collected dust is poured back onto a takeaway belt conveyor or, optionally, into big bags.

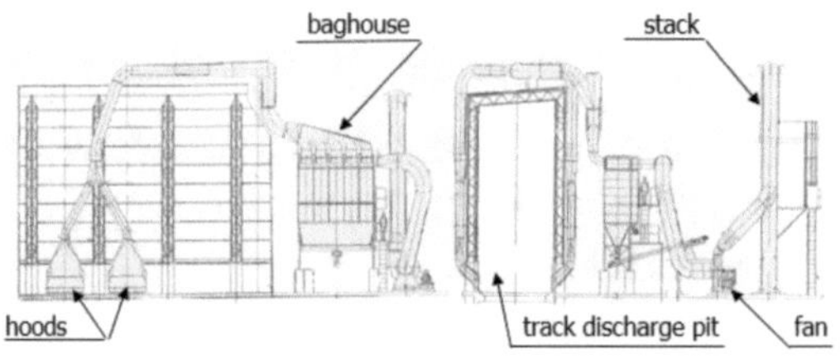

Figure 243. The discharge pit used for discharge of double tipper trucks.

Figure 244. Discharge pit for double tipper trucks, shown in Figure 243.

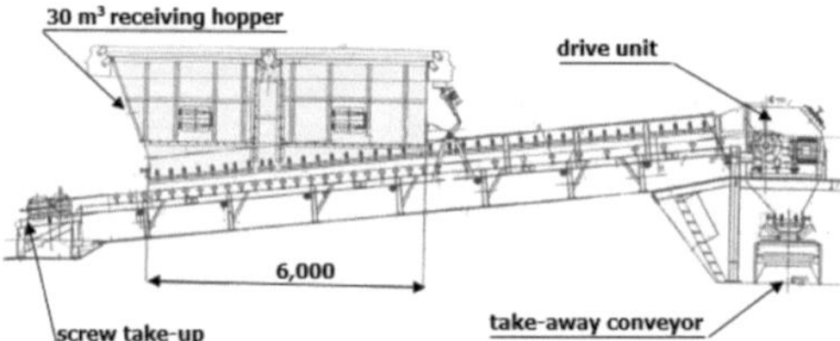

Figure 245. Belt feeder of 750 tph capacity installed in discharge the pit.

Figure 246. Discharge pit with belt feeder transferring bulk urine into chain bucket elevator feeding overhead conveyor of the storage (Finland).

4.2.2 Unloading of Non-Tipper Trucks

A non-tipper truck, also called a box truck, is shown in Figure 247. The truck is equipped with a hinged rear door that opens at a certain inclination of the truck pouring a bulk material.

Non-tipper trucks are discharged by special truck tipplers that tilt the trucks at an angle when the bulk material begins to slide downward into receiving hopper.

The use of tipplers is cost-effective only in cases where considerable number of box trucks should be discharged quickly and dustless.

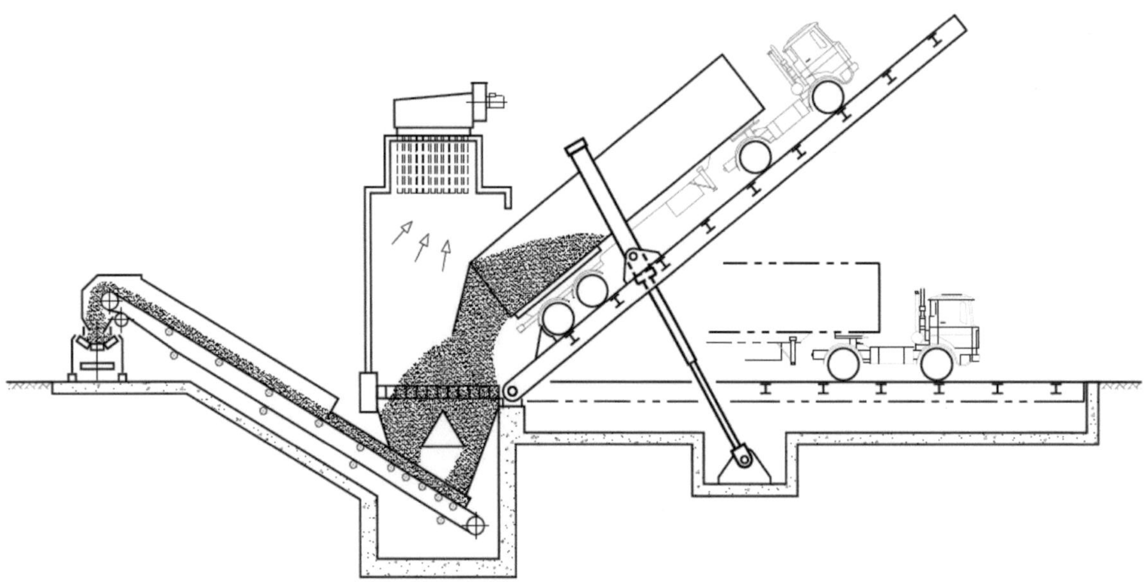

Figure 247. Scheme of hydraulic tippler for discharge of non-tipper bulk trucks.

To discharge dusty bulk material, the tippler must be equipped with a dedusting system.

In exceptional cases, covered box trucks, carrying powders, can be discharged by a vacuum pneumatic system.

4.2.3 Discharge of Tank-type Trucks

Tank-type trucks (Figure 248) are used for transport of powders and easy-fluidized bulk materials such as cement, fine phosphate, flour, so on. Tank-type trucks are discharged by dense phase pneumatic conveying. Compressed air is supplied to the fluidizing pads or pipes installed within the truck body. The fluidized fine material slides downward to the outlet pipe, where additional compressed air is introduced into the pipe in order to continue the movement of the air-material mixture to the receiving bin.

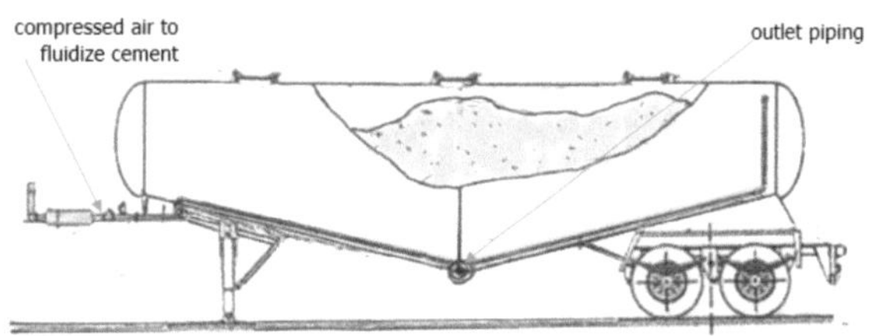

Figure 248. Scheme of tank-type cement carrier truck.

4.3 Loading of Railway Wagons

Two main types of wagons for carriage of bulk materials are in use today:

a. open wagons

b. covered bottom-dumping wagons (Figure 249).

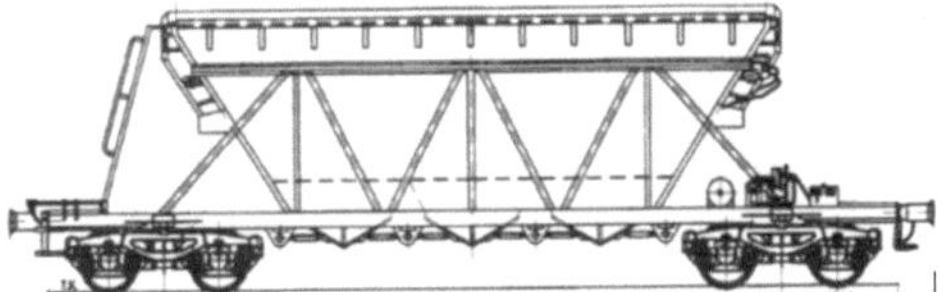

Figure 249. Covered bottom-dumping wagon with a payload of 65 ton of potash.

Open railway wagons are designed to transport bulk materials such as coal, gravel, ore, and so on, that are not affected by rains, snow, or winds. Bulk materials that can be affected by the environment (fertilizers, phosphate, salt, and so on) are usually transported in covered bottom-dumping railway wagons.

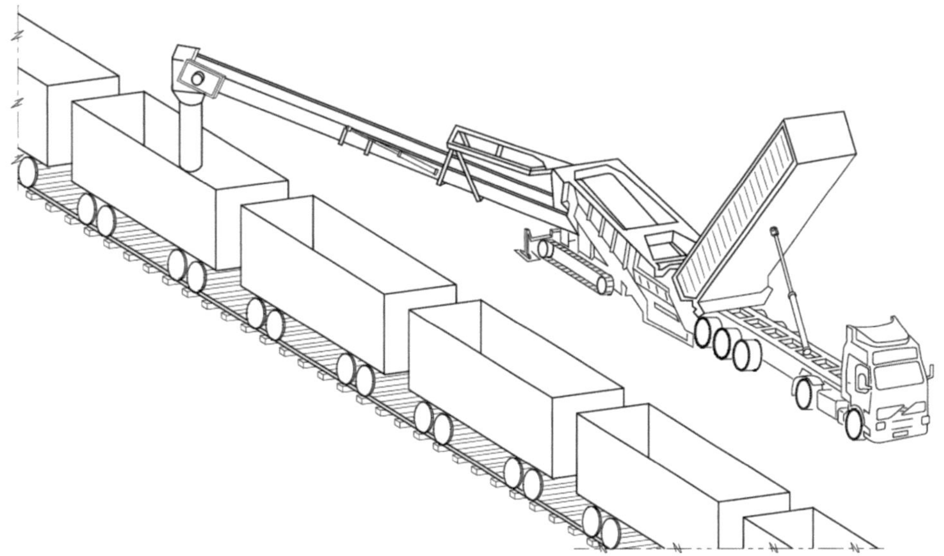

Figure 250. The loading of open wagons with mobile swivel belt conveyor.

The wagon-loading facility, shown in Figure 250, is a mobile swivel belt conveyor attached to a tipper truck discharger (Samson or similar). The train of open wagons is slowly moved by locomotive to fill the open wagons evenly, or the loading system can move (or moved) along unmoving wagons to fill one wagon after another.

To carry out the environmentally friendly loading of a dusty bulk material, the truck discharger must be equipped with a dedusting system and wagon loading chute must be the telescopic and dedusting.

Advantages of truck/wagon loading shown in Figure 250:

1. The loading does not require any special preparations on site and can begin immediately after the truck discharger is coming.
2. Low Opex and low Capex.

Disadvantages:

1. The wagon loading capacity depends on the tipper truck arrivals and on the capacity of the discharger. The average loading capacity of such system is between 100 tph and 200 tph.
2. There is no buffer storage, so, the facility operates according to the following arrangement: no trucks, no loading, and many trucks, lengthy line of waiting trucks.
3. Dust can be generated during loading of wagons.
4. No scale to weigh wagon during filling.
5. Distribution of bulk material along a wagon is a time-consuming operation.
6. Double trucks should be splitted and discharged separately.
7. A significant open area is required for the manoeuvring of trucks and for the line of trucks waiting for discharge.

Large freight turnover requires special covered, environmentally friendly and high-capacity wagon-filling facilities.

In principle, the loading facilities for wagons are similar to the trucks loading facilities and can be divided into three main groups:

Group 1

The empty wagon is mounted under the loading silo on the ground-level scale. The loading is carried out straight from the silo and is controlled by means of two scales weighing each bogie of the wagon separately. It helps to control the distribution of the load within the wagon. The silo is filled by an inclined (or horizontal) overhead belt conveyor (Figure 251).

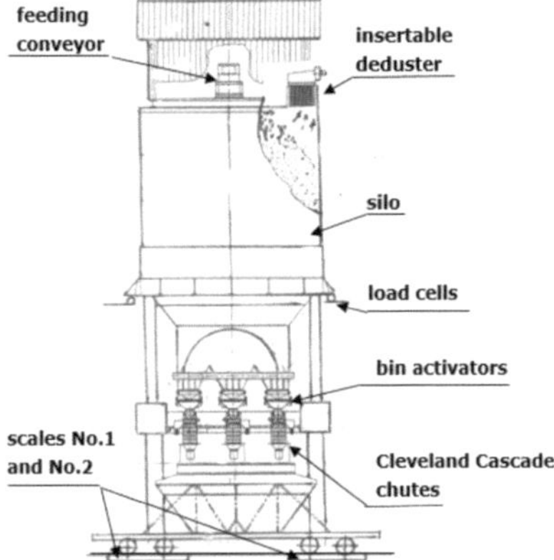

Figure 251. Scheme of the loading of the wagon from a top silo (Group 1).

Advantages of the loading system:

1. Simple loading system.
2. Three chutes provide high loading capacity (300 tph ÷ 350 tph per chute), as the nominal capacity of a silo-feeding belt conveyor must be 1,200 tph to 1,300 tph.
3. Bin activators allow rapid cleaning of the silo when is a need to replace one bulk material to another or to activate non-free-flowing bulk materials.
4. It is recommended to equip the loading system with additional diverters and bypass chutes to feed additional takeaway belt conveyor transporting the material from the loading silo back to the storage. The arrangement facilitates cleaning of the silo between different bulk materials or to transport the material from a truck passage (serving as a reserve discharge pit for tipper trucks) back to a storage.

Disadvantages:

1. The loading tower must be equipped with a dedusting system.
2. High Opex and high Capex.

Group 2

The wagon-loading operation is based on a preliminary preparing of the batch of bulk material (equal to the capacity of one wagon) into the weighed loading silo. The silo is filled up from the top surcharge bin (Figure 252).

For example, Bateman (RSA) developed a fully automatic non-stop coal-loading system for trains of open wagons, consisting of two loading silos. The wagons are slowly moved under the silos as the bottom discharge gate of the first silo opens and the wagon is loaded with prepared batch of material. Concurrently, the second silo is filled from the top bin to be ready to load the second wagon.

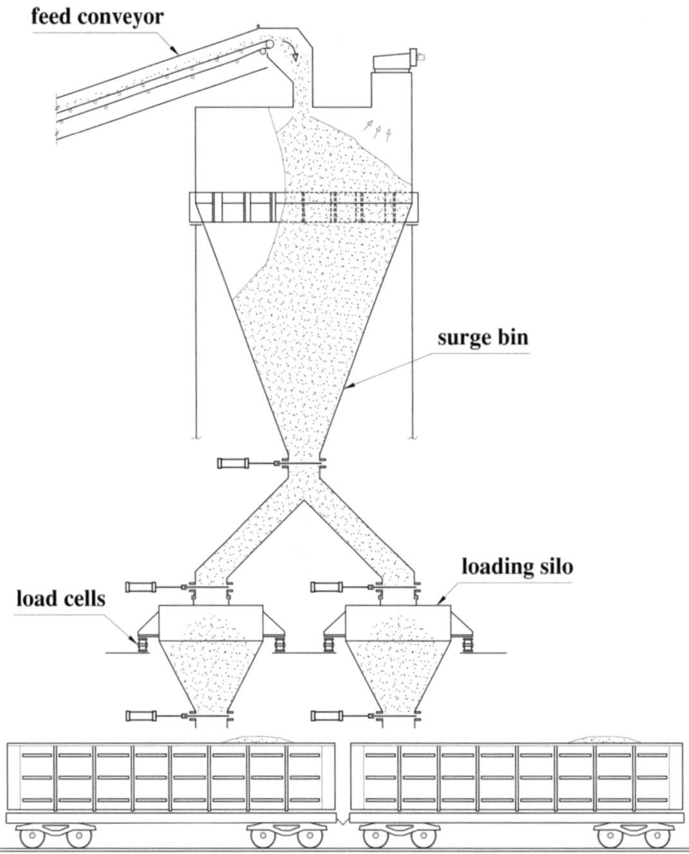

Figure 252. Loading of open wagons with the prepared batch of the bulk material (Group 2).

Advantages of the loading facility:

1. Rapid filling and discharge of the loading silo.
2. Two loading silos operating successively, allow practically non-stop loading of the slow-moving train of open wagons.
3. Accurate weighing of the batch of the payload material.
4. Ground-level wagon scales are not required.

Disadvantages of the loading facility:

1. The loading silo must be designed to be emptied completely and rapidly, as ratholing or bridging within the silo can stop the wagon loading operation.

2. The surcharge bin(s) should feed the loading silo(s) at high and constant capacity.

3. High tower located above the loaded wagons significantly increases the cost of project.

Group 3

A horizontal, low-speed belt feeder (shuttle) is slowly moved back and forth above and along the wagon opening, and evenly distributes the material along the wagon. The shuttle can be fed by a fixed belt conveyor or through the loading silo located above the feeder (Figure 253). In order to control the wagon loading, a belt scale can be installed on the feeding conveyor, or the loaded wagon can be mounted on the ground-level scale.

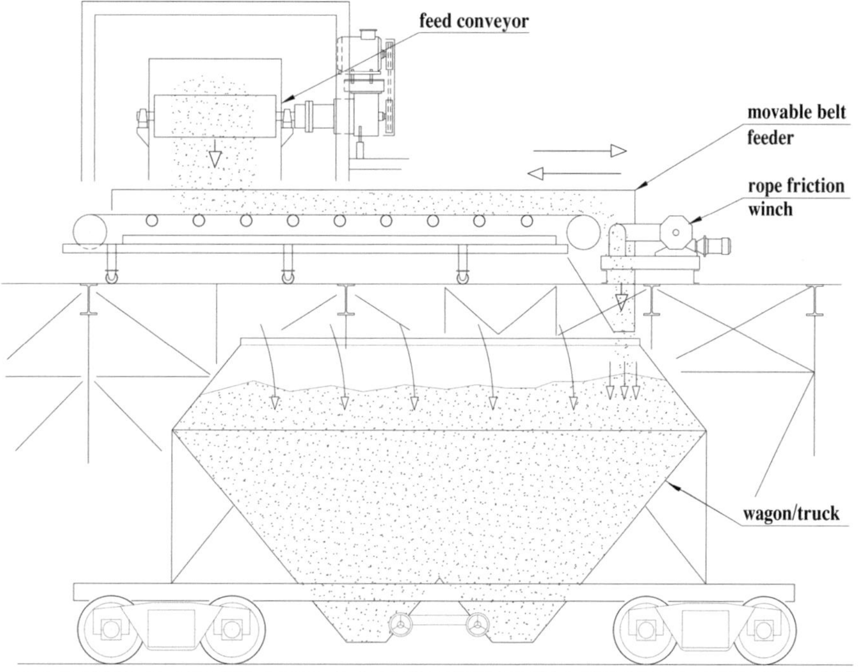

Fig. 253 Loading of a wagon by the movable (shuttle) belt feeder (Group 3).

Advantages of the loading system:

1. The loading of a wagon is carried out without an intermediate silo (buffer) thus there are no bridging or ratholing problems.

2. Uniform distribution of the loaded material along the wagon.

3. High loading capacity.

4. Material falls from a low height, so low dust emission.

5. A belt scale installed on a feeding conveyor provides high accuracy of loading.

6. A high and expensive load tower is not required.

Disadvantages:

1. 1.Opposite to no.1 of the advantages: without an intermediate silo (buffer), the feeding conveyor should operate nonstop. Any stop of the feeding conveyor causes concurrent stop of the wagon/truck loading operation.

2. A special CCTV camera is recommended to be installed for visual control of the loading process.

4.4 Discharge of Wagons

The discharge principles for covered bottom-dumping wagons and for open wagons are different.

4.4.1 Discharge Pits for Covered Wagons

Covered bottom-dumping wagons are discharged within special enclosed and dedusted discharge pits, where they drop the bulk material into underground receiving hopper. The store capacity of the receiving hopper should be 1.3 to 1.5 times the capacity of the loaded wagon. This requirement is justified because discharge of such wagons is very fast. For example, the wagon equipped with Ortner (USA) bottom-dumping doors (Figure 254) drops 65 tons of potash in 8 to 10 seconds, and the Arbel (France) 65-ton wagons are emptied between 10 and 15 seconds. The hopper should be big enough to receive the wagon payload without piling the bulk material upward from the pit grizzly because this pile will interfere with the movement of the train and with positioning of the next loaded wagon over the hopper.

Typical wagon discharge pits for bulk materials are shown in Figure 255 and Figure 256. The receiving hoppers of the discharge pits are commonly discharged by belt feeders (drag chain feeders are used only in exceptional cases). The covered pit is usually provided with a dry dedusting system that can be installed on the roof of the pit or located close to the pit.

Figure 254. Discharge doors mounted on potash bottom-dumping wagon (Ortner Co., USA).

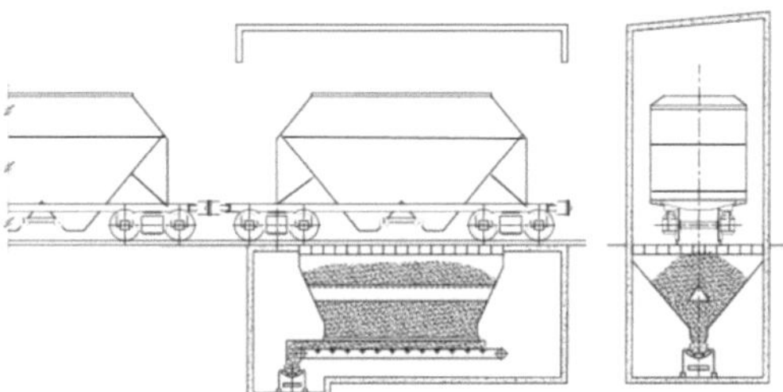

Figure 255. Wagon discharge pit with hopper with belt feeder.

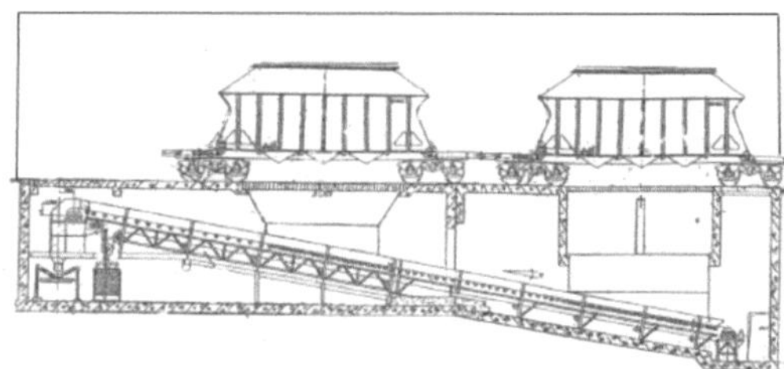

Figure 256. Discharge pit for two bottom-dumping wagons.

The belt feeder is the optimal solution for the most of discharge pits because of its simplicity, reliability, low-maintenance requirements, and capability to reach any required discharge capacity at the lowest cost.

To prevent impact on the feeder caused by the mass of material falling from the wagon, and to reduce dust emission, special deflectors can be installed within the receiving hoppers. Optionally, the deflection plates can be installed in a zigzag pattern (the principle of Cleveland Cascade chutes) within the hopper to reduce the speed of the falling bulk material.

4.4.2 Rotary Tipplers

Open wagons are used for large fright turnarounds and, thus, require special unloading terminals.

For example, coal power stations use a huge amount of bulk coal that is transported in long trains of open wagons.

To discharge a train of open wagons without separating them, special rotating machines (tipplers) are developed and widely used.

A rotary tippler with receiving rated load of 120 tons of coal and at a maximum tipping angle of 175°, can provide 1.0 min to 1.5 min discharge time per wagon.

The tippler automatically fastens each entering and positioned loaded open wagon, rotates it, pours the bulk material out into the receiving hopper, and rotates the empty wagon back to its initial position.

Then this empty wagon is automatically unfastened and moved out of the tippler, as a new loaded wagon (coupled with the previous wagon and also with the following one) is set in discharge position, ready to be fasten and to be unloaded during rotation. The cycle is repeated.

Figure 257 shows the discharge of open coal wagons in a Metso tippler.

Figure 257. Unloading an open wagon in a tippler (Metso).

CHAPTER 5

Shiploaders and Ship Unloaders

SHIPLOADERS

General

Millions and millions of tons of various bulk materials are delivered every year to the export seaports by rail or by truck. The bulk materials are unloaded in port discharge pits and stored in storages (or direct handled), occasionally scalped or screened, and loaded into vessels or into barges by various shiploaders.

A shiploader is a large machine used for continuous loading of bulk materials into vessels or barges. Each shiploader is a custom-engineered machine ordered and manufactured to suit a customer's specific requirements, such as:

- type of load-out wharf or jetty
- bulk material characteristics
- required loading capacity
- type of vessels to be loaded
- local and international environmental rules and regulations
- budget limitation

A typical shiploader consists of a portal (movable or fixed); an upper structure (A-type or other); a swivel or non-swivel boom, pivotally connected to the portal or to the upper structure; a fixed or shuttle-type boom belt conveyor; a boom tilting winch (or tilting hydraulic cylinders); a telescopic dedusting loading chute, connected straight (or via an additional tilting chute) to the head chute of the boom conveyor; a power/control cable reeling drum, and so on.

There are so many different types of shiploaders (Figure 257) that the most important task for the engineer is to choose the type of shiploader which will be the right solution for the given conditions. This optimal type of shiploader, chosen by the engineer, will be the basis for the tender for the supply of a new shiploader.

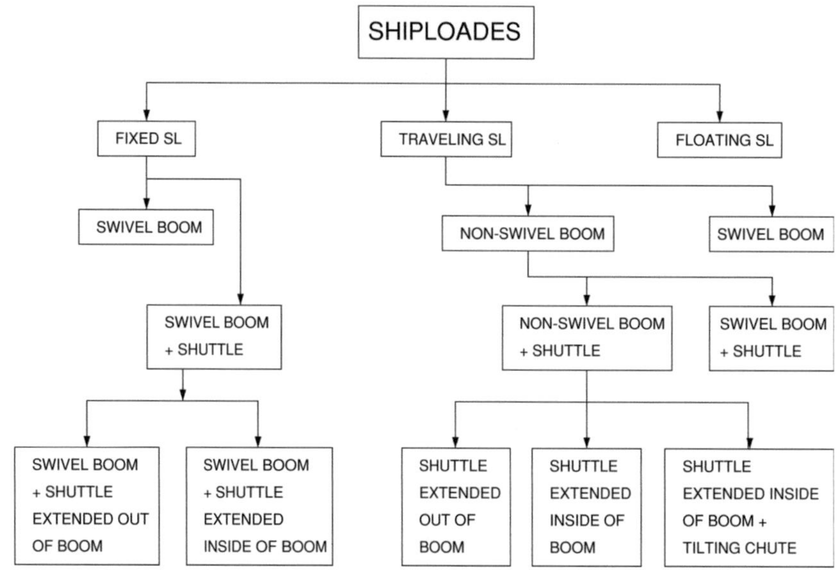

Figure 257. Various types of continuous shiploaders for bulk materials.

5.1 General Description of Shiploading Operations

The loading of vessels by a shiploader is subjected to certain rules and regulations. In order not to overstress the hull of the vessel, the master of the vessel and the terminal representative should agree on a loading plan (cargo plan, below) and should sign the plan as agreed before commencing the shiploading operation (S.I. No. 347/2003).

There are two sets of permissible loads on loaded vessels: still water shear forces (SWSF) and steel water bending moment (SWBM). One set is for seagoing vessels (at sea), and the second is for vessels in harbour (during loading in port) (Figure 258).

In port, the hull girder is permitted to carry a higher level of stress imposed by the static load.

In any case, the SWSF and SWBM of the vessel are not to be exceeded.

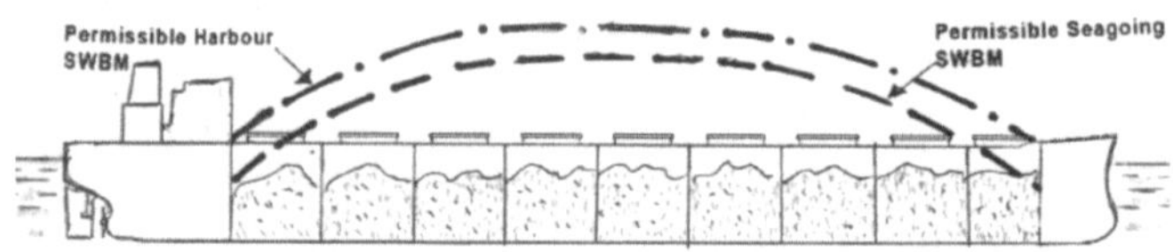

Figure 258. Permissible bending moments (port and seagoing) of the loaded vessel.

The vessel officer in charge should ensure that:

- The agreed loading/unloading sequence is being followed by the terminal.
- Any damage to the ship is reported.

- The cargo is loaded, where possible, symmetrically in each hold and, where necessary, trimmed.
- Effective communication with the terminal is maintained.
- The terminal staff advise of pour completions and movement of shoreside equipment in accordance with the agreed plan.
- The loading rate does not increase beyond the agreed rate for the loading plan.

An example of vessel cargo plan (sequence of holds loading) is shown below:

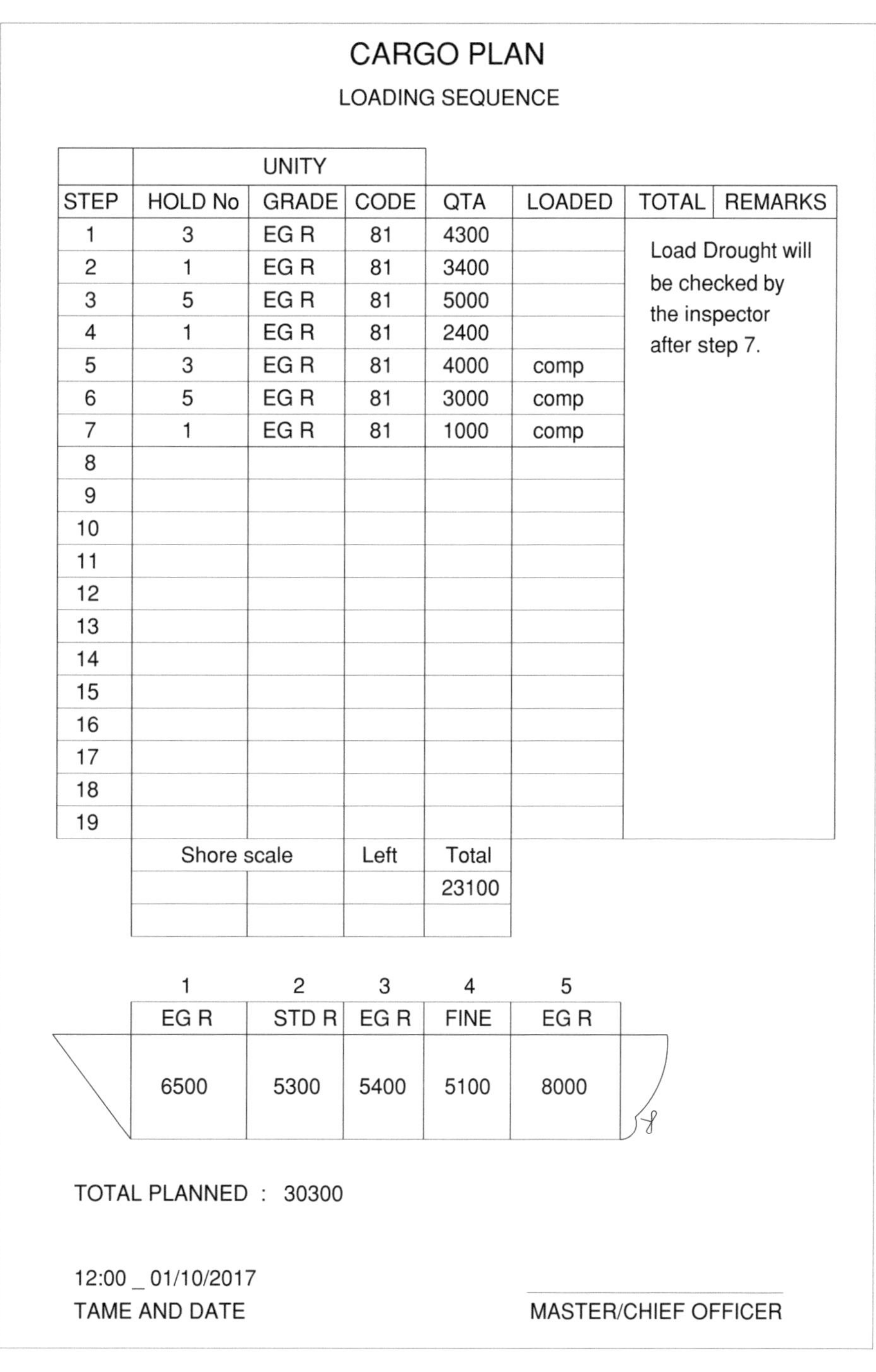

CARGO PLAN

LOADING SEQUENCE

	UNITY						
STEP	HOLD No	GRADE	CODE	QTA	LOADED	TOTAL	REMARKS
1	3	EG R	81	4300			Load Drought will be checked by the inspector after step 7.
2	1	EG R	81	3400			
3	5	EG R	81	5000			
4	1	EG R	81	2400			
5	3	EG R	81	4000	comp		
6	5	EG R	81	3000	comp		
7	1	EG R	81	1000	comp		
8							
9							
10							
11							
12							
13							
14							
15							
16							
17							
18							
19							

Shore scale		Left	Total
			23100

1	2	3	4	5
EG R	STD R	EG R	FINE	EG R
6500	5300	5400	5100	8000

TOTAL PLANNED : 30300

12:00 _ 01/10/2017
TAME AND DATE

MASTER/CHIEF OFFICER

Figure 259. Example of cargo plan.

According to 13-98-IMO [38], the following statement between the representative of the loading terminal and the master of the bulk carrier should be agreed to:

> "The terminal representative is responsible for loading with the hatch sequence and tonnages stated on the ship's loading plan. In addition, the terminal representative should:
>
> - Complete the checklist in consultation with the master before the loading is commenced.
> - Not deviate from the loading plan unless by prior consultation and agreement with the master.
> - Trim the cargo, when loading, to master's requirement.
> - Maintain a record of the weight and disposition of the cargo to ensure that the weights in the hold do not deviate from the plan.
> - Provide the master with names and procedures for contacting the terminal personnel or shipper's agent who will have responsibility for the loading operation and with whom the master will have contact.
> - Avoid damage to the ship by loading equipment and inform the master if damage occurs.
> - Ensure that hot works is carried out on board or in the vicinity of the ship while the ship is alongside the berth except with the permission of the master and in accordance with any requirements of the port administration.
> - Ensure that there is agreement between the master and representative at all stages and in relation to all aspects of the loading operation".

To prevent the migration of slope failure of material in hold during the cargo ship floating, the IMO (2009) requires that bulk materials with a repose angle of 30° or less must be trimmed during the loading (Figure 260).

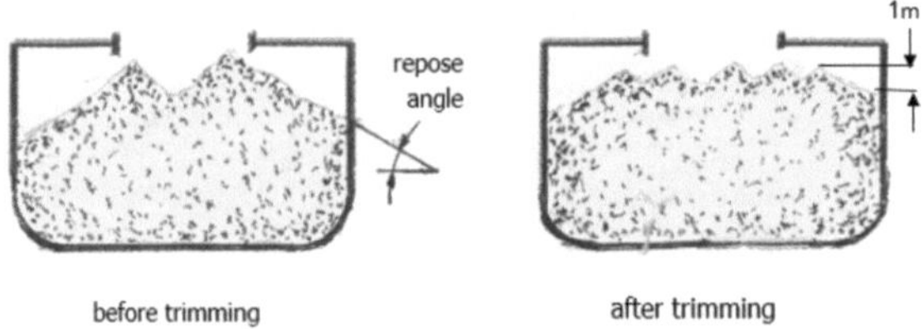

Figure 260. Requirement for trimming as a function of repose angle of the bulk material.

5.1.1 Stability of Shiploaders

To check the safety of the shiploader against overturning [39], the stability ratio or safety factor ($v_k = M_s/M_k$) must be calculated depending on the prescribed load combinations (see below).

Table 4

Sub-clause	Type of load	Main loads	Main and additional loads	Main additional and special loads								Erection load
		I	II	III 1	III 2	III 3	III 4	III 5	III[1)] 6	III 7	III 8	III 9
3.1.1	Dead loads	x	x	x	x	x	x	x	x	x	x	x
3.1.2	Material loads on conveyors, reclaiming divices and hoppers	x	x	x	x	x	x	x	x	x	x	
3.1.3	Incrustation	x	x	x	x	x	x	x	x	x	x	
3.1.4	Normal digging and lateral resistances	x		x		x		x				
3.1.5	Forces on the conveyor	x	x	x	x	x	x	x	x	x	x	
3.1.6	Permanent dynamic effects	x	x	x	x	x	x	x		x	x	
3.1.7	Loads due to inclination of machine	x	x	x	x	x	x	x	x	x	x	
3.2.1	Wind load during operation [2)]		x	x	x	x	x	x		x	x	
3.2.2	Snow and ice (possibly)											
3.2.3	Temperature (possibly)											
3.2.4	Abnormal digging and lateral resistances		x									
3.2.5	Resistances due to friction and travel		x									
3.2.6	Reactions pependicular to the rail		x									
3.2.7	Non-permanent dynamic effects		x									
3.3.1	Blockage of chutes			x								
3.3.2	Bucket-wheel resting				x							
3.3.3	Failure of safety devices					x						
3.3.4	Locking of travelling device						x					
3.3.5	lateral collision with the slope (bucket wheel)							x				
3.3.6	Wind load on non operating machine								x			x
3.3.7	Buffer effects									x		
3.3.8	Loads due to eartnquakes										x	
3.3.8	Loads due to eartnquakes											x
3.3.9	Erection loads (deed loads in particular situations)											

1) The removal of abnormal digging resistances (see 3.2.4) shall be ensured, when necessary, by appropriate devices (locking device which prevents slewing of appliance when out of service due to wind force).

2) See 3.2.7

Table 31. **Safety against overturning**

$$v_k = M_s / M_k$$

Load case	v_k
I	**1.5**
II	**1.3**
III	**1.2**

where

M_s is the stabilizating moment

M_k is the overturning moment

The stabilizing moment to overturning moment ratio (safety factor V_k) is to be calculated relative to the tipping axis *A* (Figure 261).

An example of simplified calculations of the stability of a shiploader is shown below:

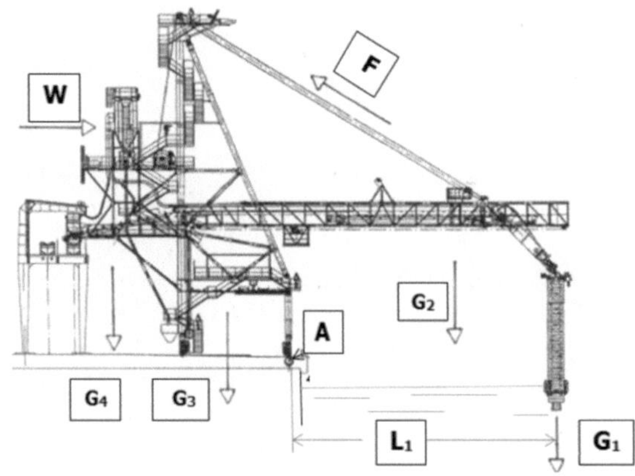

Figure 261. Example of simplified calculation of shiploader stability.

$$V_k = M_s/M_k = \frac{G3 \times L3 + G4 \times L4 + F \times Lf}{G1 \times L1 + G2 \times L2 + W \times Lw}$$

where V_k is the safety factor; M_s is the stabilizing moment; M_k is the overturning moment; G_1 is the weight of the Cleveland Cascade chute blocked by the bulk material; G_3 is the weight of the shiploader portal and A-frame; G_2 is the weight of the boom with loaded conveyor when the shuttle is fully extracted; G_4 is the weight of the shiploader tail section; *W* is the wind overturning force; and *F* is the tension of the luffing winch steel ropes.

Booms of the shiploaders are supported by two separate wire ropes wound on the drum of the luffing winch.

In accordance with Euler's law, four turns of each rope on the winch drum (plus fastening the end of the rope with clamps) create a frictional force sufficient that the rope does not slip off the drum at the *maximum permitted negative angle* of the boom. Increasing the boom tilt down by a negative angle greater than the permitted angle can cause the boom to fall.

The safety factor *(SF)* of the wire ropes must ensure safety of the boom even in the case of the breakage of one of the ropes. The total safety factor of ropes is $SF = 2 \times SF_1 = 6$, and, after failure of one of the ropes, the safety factor will be $SF' = SF_1 = 3$.

1.1.2. Safety Considerations.

At majority of shiploaders, the boom is supported, raised and lowered by means of a rope luffing winch. The typical winch consists of a rope drum, a reducer, an electric motor and brakes.

The brakes are one of the most essential elements, providing safety of the shiploader operation. The operational or service brake is usually electro-hydraulic thruster drum brake installed between the motor and the gearbox (sometimes using two operational brakes, installed on both sides of the gearbox) while the emergency brake is a calliper disk, spring applied and hydraulic pressure released brake installed on the rope drum of the luffing winch.

There are two options for operation of the operational and the emergency brakes:

Option no.1

1. Operational brake is a "normally closed" brake that opens concurrently with the supply of electricity to the *motor* and is closed at the same time as the power supply to the motor is cut off.
2. Emergency brake is a "normally closed" brake opens concurrently with the supply of electricity to the *shiploader* and is closed when supply of electricity to the *shiploader* is cut off. During shiploading, the brake is in the open, stand-by state and is trigged only by receiving a signal from the winch drum rotational speed sensor (for example, the speed exceeds the rated speed by 3%÷5%) or from slack rope sensor.

Option no.2

1. Operational brake is a "normally closed" brake, that opens concurrently with the supply of electricity to the *motor* and is closed at the same time as the power supply to the *motor* is cut off.
2. Emergency brake is a "normally closed" brake that opens each time *before* the operational brake opens (delta time is 1.0÷1.5 sec) and is closed *after* the operational brake is closed with the same time difference, and all this in addition to the trigging by receiving a signal from the drum rotational speed sensor or from slack rope sensor.

Each of these options has supporters and opponents (we have been successfully using option no.2), but under all conditions, the brakes require regular inspection, maintenance, strict implementation of all the instructions of the brake manufacturer. It is recommended to equip brakes with excessive pad wear limit-switches.

Figure 261.1 shows what happened when all three (two operational and one emergency) brakes of the shiploader were out of order.

Figure 261.1. It happened when out-of-order brakes did not prevent free rotation of the luffing winch rope drum, ended with the fall of the shiploader boom on the vessel (May 2018).

5.2 Fixed Shiploaders

A fixed shiploader is an immovable machine installed on a fixed foundation on a wharf or offshore on a piles-supported platform.

The fixed shiploaders are equipped with a slewing boom with fixed conveyor or with slewing boom with shuttle-type conveyor. A dedusting telescoping chute is suspended from the lower section of the head chute of boom conveyor.

The fixed shiploader is fed by a fixed belt conveyor.

The fixed slewing boom shiploader shown in Figure 262 was manufactured by TAKRAF Tenova (Antamina, copper/zinc concentrate, at shiploading capacity of 1400 tph).

Figure 262. The fixed shiploader with a slewing boom and shuttle boom conveyor extended out of the boom structure (TAKRAF Tenova).

The shiploader is composed of a central stationary column, a swivel boom with a shuttle-type conveyor extending out of the boom structure, a slewing mechanism, a dedusting telescopic loading chute, and hydraulic cylinders for the tilting of the boom. The shiploader is fed by the fixed slope conveyor through a chute as the centre-line of the chute coincides with the centre-line of rotation of the boom.

The row of cranes installed on the deck of the typical bulk carrier (Figure 263) cause the serious operational difficulties for operation of a fixed shiploader and its inability to reach every point of every hold without multiple stops of loading and shifting of a large vessel.

Advantages of a fixed shiploader:

1. The long wharf, that can withstand the heavy load of a moving shiploader, is not needed, it is required that this is inexpensive fixed foundation for the shiploader.
2. No wharf bridge with feed conveyor and tripper are required.
3. Barges (1,000 dwt÷3,000 dwt) and small bulk carriers can be loaded without shifting.

Disadvantages of a fixed shiploader:

1. Today most bulk carriers are equipped with rows of cranes (Figure 263), so the installation of the boom of a fixed shiploader between the cranes for the loading and trimming is the extremely hard and time-consuming operation.
2. The slewing and shuttle motions of the boom of a fixed shiploader allow to reach a relatively small loading area between cranes, so large vessels should be shifted along the wharf several times to be loaded and trimmed (in accordance to the cargo plan). The loading operations are stopped during the shifting, so the average loading capacity is reduced significantly. In addition, the frequent shifting of the vessel requires non-stop operation of the vessel crew and the facility personnel.

3. Control over a fixed swivel-boom shiploader operation (from the deck and/or from the shiploader control room) is much more complicated than control over travelling shiploaders.
4. The cost of a fixed shiploader is practically the same as the cost of a travelling shiploader.

Figure 263. Typical bulk carrier is ready for loading.

Figure 264. Three fixed swivel and shuttle-type boom shiploaders connected by fixed belt conveyors ("three-towers" solution), Bedeschi, Italy.

In order to overcome the main disadvantages of fixed shiploaders, an innovative solution was proposed: the "three-towers" system (three fixed swivel-boom shiploaders with shuttle-type boom conveyors were installed in a row, Figure 267, Figure 268). This loading system, designed and manufactured by Bedeschi (Italy), allows vessels of 50,000 dwt÷180,000 dwt to be loaded with coal without shifting at a capacity of up to 6,000 tph.

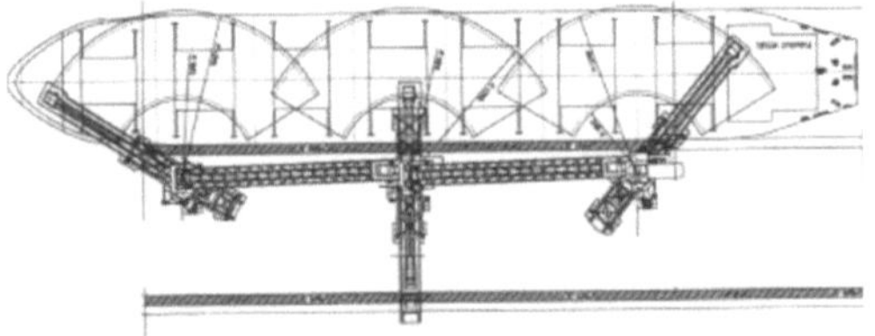

Figure 265. The layout of the "three towers system", showing the loading of a Panamax vessel without shifting (Bedeschi, Italy).

5.3 Travelling Shiploaders with Swivel Booms

A travelling swivel boom shiploader consists of a portal, mounted on the motorized bogies and moving on rails; an upper slewing structure (A-type); a boom structure pivotally connected to the slewing upper structure; a boom conveyor; a telescopic loading chute; and so forth.

These shiploaders travel along the wharf on two parallel rails. The distance between the rails varies from 10 m to 15 m.

Depending on the wharf structure, the sea-side rail is located 2 m÷5 m from the edge of the wharf.

According to the regulations, while traveling the shiploader, no part of the shiploader should hang over the line of the wharf, so the boom is to be tilted upward or slewed along the wharf.

The wharf gallery (bridge) is installed on and along the wharf and carries a belt conveyor, delivering bulk material from storages and/or from discharge pits.

The wharf belt conveyor is discharged by a tripper, that mechanically connected to the tail section of shiploader intermediate conveyor. The tripper is towed by the shiploader transferring the material to the intermediate belt conveyor and further to the boom conveyor.

A typical travelling shiploader with a slewing boom and belt conveyor, fixed inside the boom structure, is shown in Figure 266 and Figure 267. This shiploader was manufactured by SMAG, Germany, in 1992.

The travelling shiploader manufactured by Bedeschi looks a little different (Figure 268).

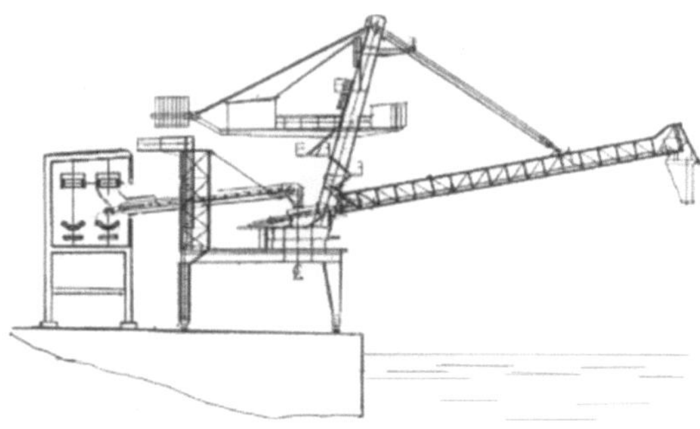

Figure 266. Scheme of travelling shiploader with swivel boom.

Figure 267. Travelling shiploader with swivel boom (SMAG, Germany).

Figure 268. Travelling shiploader with swivel boom (Bedeschi, Italy).

Two shiploaders (Figure 267 and Fig. 268) are distinguished by a system that provide the inclination of the boom. The boom of SMAG shiploader is suspended by steel wire ropes to the luffing winch that allows the boom to be tilted in the range from +65° (to be locked) to -10°. This arrangement requires the lengthening of the tail section of the boom conveyor and separation of the centre-line of the boom rotation from the centre-line of the boom tilting. This separation often causes belt dislodgement problems, especially when the boom is tilted to higher angles.

The counterweight was installed on the top section of the L-shaped structure of the shiploader.

In contrast to the SMAG shiploader, the boom of Bedeschi shiploader is supported by hydraulic cylinders, allowing the boom tilting range from +40° to –5°. The centre-line of the boom rotation is combined with the centre-line of the boom tilting, thus there is no problem with the belt dislodgement.

Both systems can operate successfully. The project engineer and the customer should make the right choice analysing all aspects of the ship loading.

The advantages and disadvantages of travelling shiploaders with swivel boom are presented below:

Advantages of travelling shiploaders with slewing boom:

1. In comparison with fixed shiploaders, the travelling movement along wharf/ vessel makes the loading operation much easier: the shiploader can reach every hatch/hold and come back for trimming without shifting of the vessel.

2. During loading of or barges or bulk carriers without cranes or other obstacles on the deck, the slewing plus the travelling movements of the shiploader allow the loading chute to reach every spot of each hold.

3. The boom can be slewed and lowered along wharf in order to simplify inspection and maintenance.

Disadvantages:

1. During loading of vessels with cranes installed on the deck (Fig. 263), the problems of loading/ trimming, especially the land-side area of the holds, become more serious and sometimes such vessels should be moved seaward from the wharf to be fully loaded.

2. The conveying system of the shiploader consists of two conveyors. The additional inclined belt conveyor (and, hence, an additional transfer point) should be installed between the bridge tripper and the tail of the boom conveyor.

3. The new transfer point does not allow the tilting of the boom to its maximum upper position, to an angle of about +65°. Besides that, the new transfer point (as any transfer point) generates dust emission.

4. This solution requires significant investment (including construction of wharf bridge gallery, bridge belt conveyor with tripper, and so on).

5.4 Travelling Shiploaders with Swivel and Shuttle-type Booms

Travelling shiploader with swivel and shuttle boom conveyor (Figure. 272 and Figure 273) presents the attempts to overcome serious operational problems of a shiploader with swivel boom, arising especially during loading of vessels with row of cranes or other obstacles on a vessel's decks.

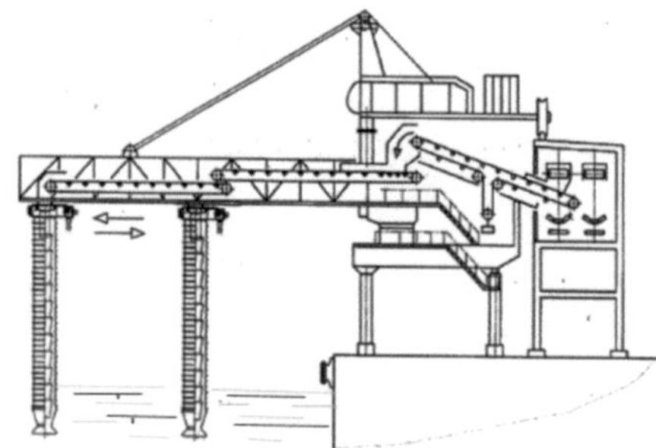

Figure 269. The A-frame shiploader with swivel and shuttle-type boom. The shuttle conveyor is extended inside the boom structure.

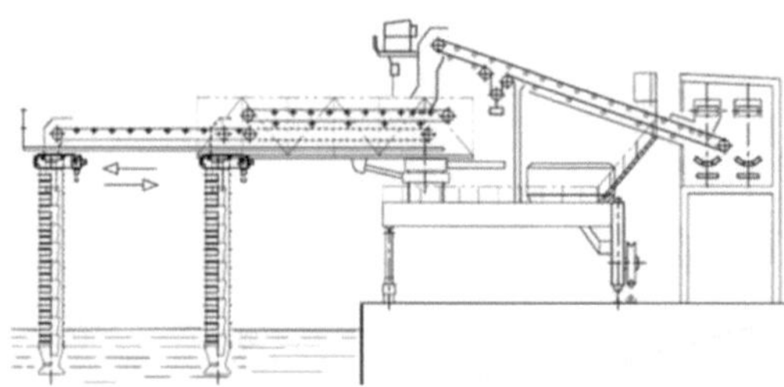

Figure 270. Shiploader (without A-frame) with swivel and shuttle-type boom. The shuttle conveyor is extended outside the boom structure.

In comparison with conventional slewing boom shiploaders, the great *advantage* of these shiploaders is the additional movement of the loading chute (shuttle movement along the boom) that simplifies loading of vessels carrying cranes on deck, but these shiploaders also bear the *disadvantages* of slewing boom shiploaders:

- additional inclined conveyor between bridge tripper and boom conveyor,
- additional transfer point with dust emission problems,
- additional slew mechanism to be maintained,
- boom can't usually be tilted more than +45° because of the transfer point limitation,
- high Opex and high Capex.

5.5 Travelling Shiploaders with Non-Swivel Booms and Shuttle-type Boom Conveyors.

There are two main modifications of the shuttle-type shiploaders:

a. Two separate belt conveyors assembled one under the other inside the boom. The upper conveyor is fixed one and fed by the tripper of the wharf conveyor to transfer the bulk material onto the lower conveyor. The lower conveyor (or shuttle), carrying a telescopic dedusting loading chute, moves (or moved) on rails forth and back. The shuttle conveyor can move inside the boom or can be extended seaward out of the boom structure.

   *Advantage of "**a**" modification:*

   Additional (shuttle) movement of the loading chute along the boom makes loading operations much more effective.

   Disadvantages:

   1. This arrangement requires boom of considerable height to install two separate belt conveyors one under the other.
   2. Two separate belt conveyors, mounted inside the boom, mean two drive units, two take-up systems equipped with manually regulated screw take- up, and so on, required regular inspection and maintenance.
   3. It is difficult to inspect, to replace idlers, pulleys and belts of "two-story"- assembled belt conveyors.

b. One telescopic belt conveyor consists of a fixed tail section that is fed by wharf conveyor through its tripper, and the telescopic head section, that is installed on the movable shuttle and can be extended or retracted inside the boom structure.

The shuttle is the special trolley that about 20-m long and weighing about 20 tons. The shuttle travels on the tracks, driven along the boom by special steel rope winch (or by rack-and-pinion steering system – the solution we can't recommend) and carries front section of the boom conveyor (together with the drive motorized pulley and the telescopic loading chute). The stroke of the shuttle traveling is usually 10 m÷15 m.

The principle scheme of a shuttle belt conveyor is shown in Figure 271.

The shuttle can move forth and back inside the boom or can be extended outside the boom structure.

*Advantages of "**b**"-modification:*

1. Additional movement of the loading chute along the boom makes the loading operations much more effective.
2. Only one belt conveyor to inspect and maintain.
3. Transfer of material from the fixed section to the shuttle section of boom conveyor does not need an additional height (Figure 271), so it is much easier to carry out inspection, maintenance, and idlers, pulleys and belt replacement.
4. The gravity take-up counterweight does not, practically, move when the tripper travels.

Disadvantages:

No serious disadvantages were observed during many years of operation.

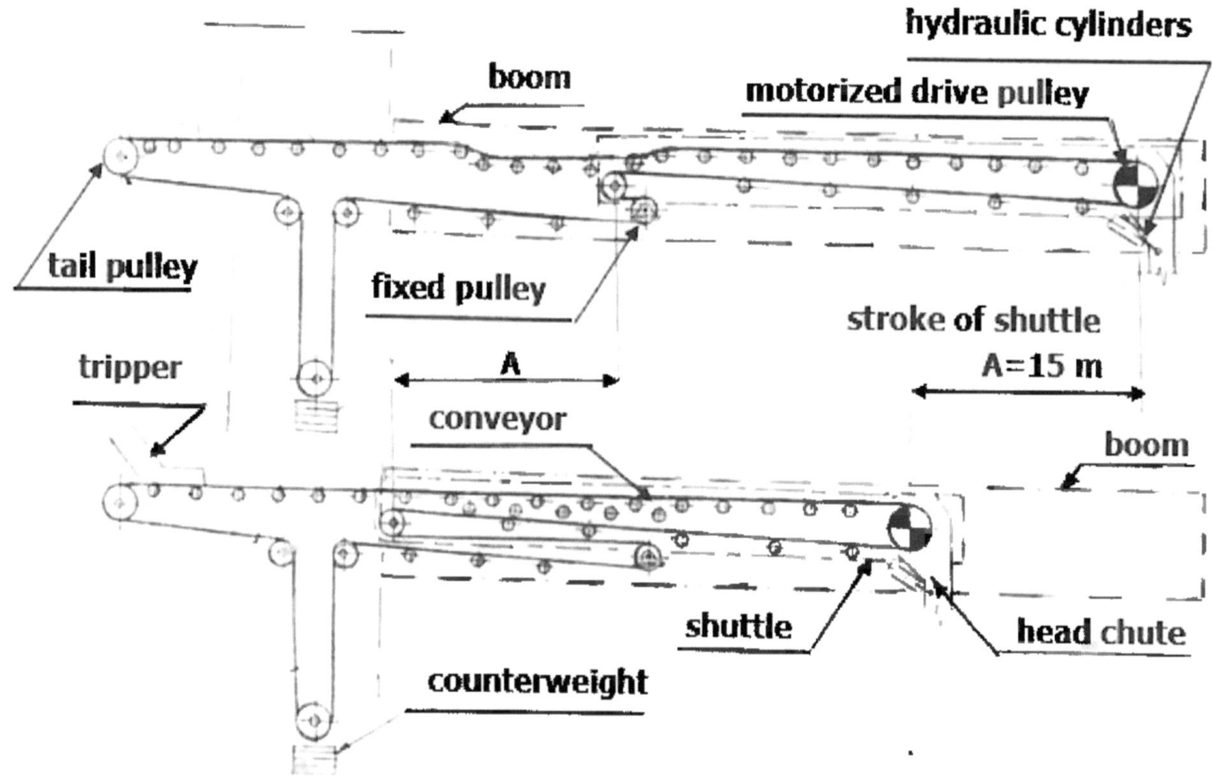

Figure 271. Scheme of the shuttle conveyor movement along the boom.

Shiploaders with a shuttle, extended and retracted inside the boom structure, are manufactured by Bedeschi (Figure 273, Figure 274, Figure 275, Figure 276), TAKRAF Tenova (Figure 272), FAM, Gapro, and others.

The shiploader, manufactured by TAKRAF for the port of Ust-Luga, Russia, is shown in Figure 280. The shiploader was designed with a shuttle conveyor extended outside the boom structure.

Figure 272. Shiploader with the shuttle moving inside the enclosed boom structure (TAKRAF Tenova).

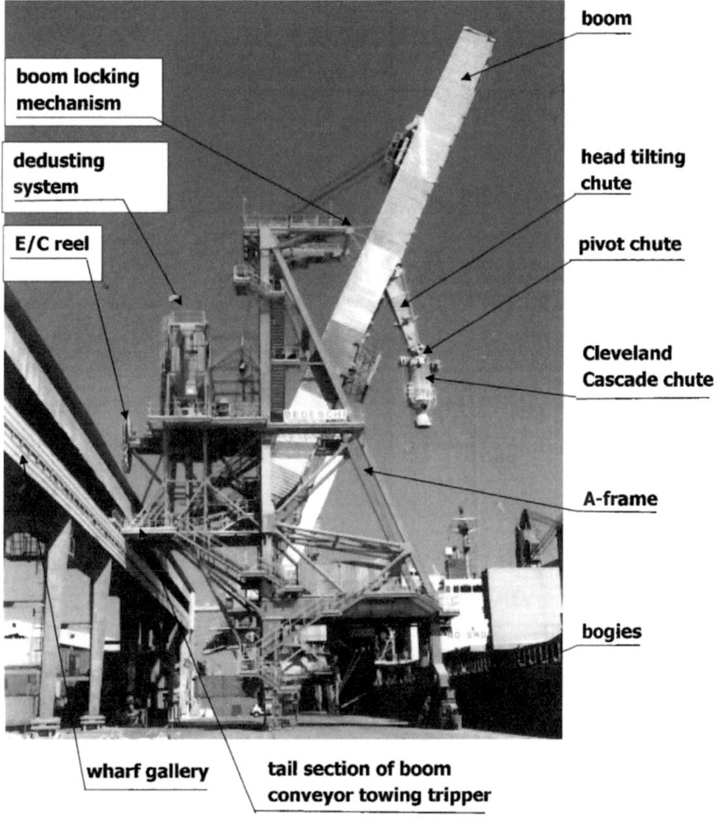

Figure 273. Shuttle-type travelling shiploader with non-swivel boom and internal shuttle at its rest position (boom is automatically locked at angle of +65°) (Bedeschi, Italy).

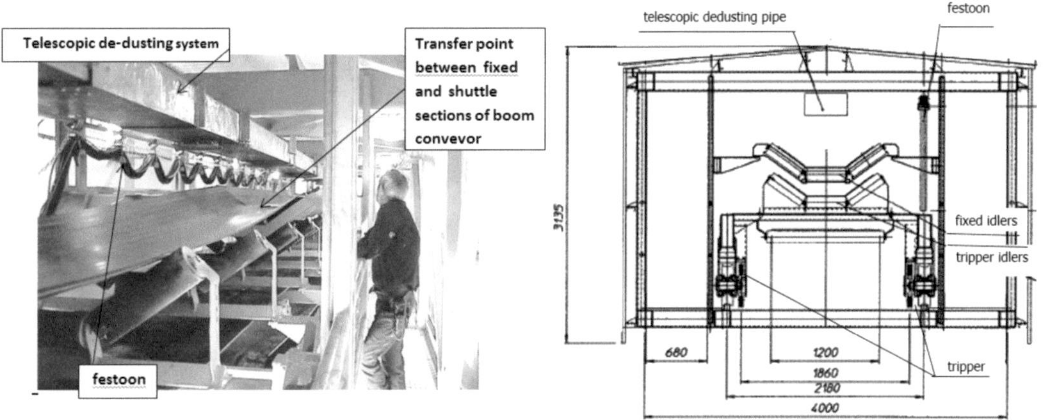

Figure 274. Enclosed boom of with internal shuttle, A, and the cross section of this boom, B, (see

Figure 273), (Bedeschi, Italy).

Figure 274 *A* is a picture taken inside the enclosed boom of a Bedeschi (Eilat) shiploader. The picture shows transfer point between the fixed section and the shuttle section of the boom conveyor, the telescopic dedusting pipe (stretched from the head chute of the boom conveyor to the fixed bag filters (2 x 10,000 Nm^3/hr) mounted on A-frame of the shiploader, the festoon suppling power/control cables to the head motorized drive pulley of the boom conveyor.

Figure 274 *B* is a cross-section of the boom.

Figure 275. Bedeschi shuttle-type shiploader operates with backward-tilted head chute and upright, extended Cleveland Cascade dedusting chute.

The Cleveland Cascade chute can be installed for maintenance and cleaning on the service platform that hydraulically moved forth and back (Figure 276).

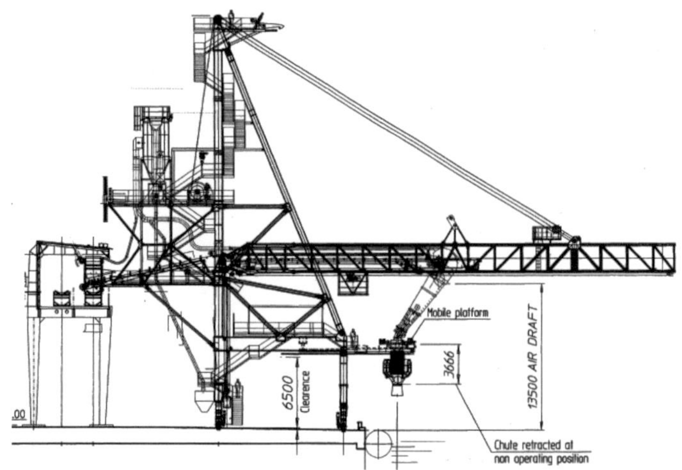

Figure 276. Shiploader with a backward-tilted head chute for the installing of Cleveland Cascade on the extended service platform (Bedeschi).

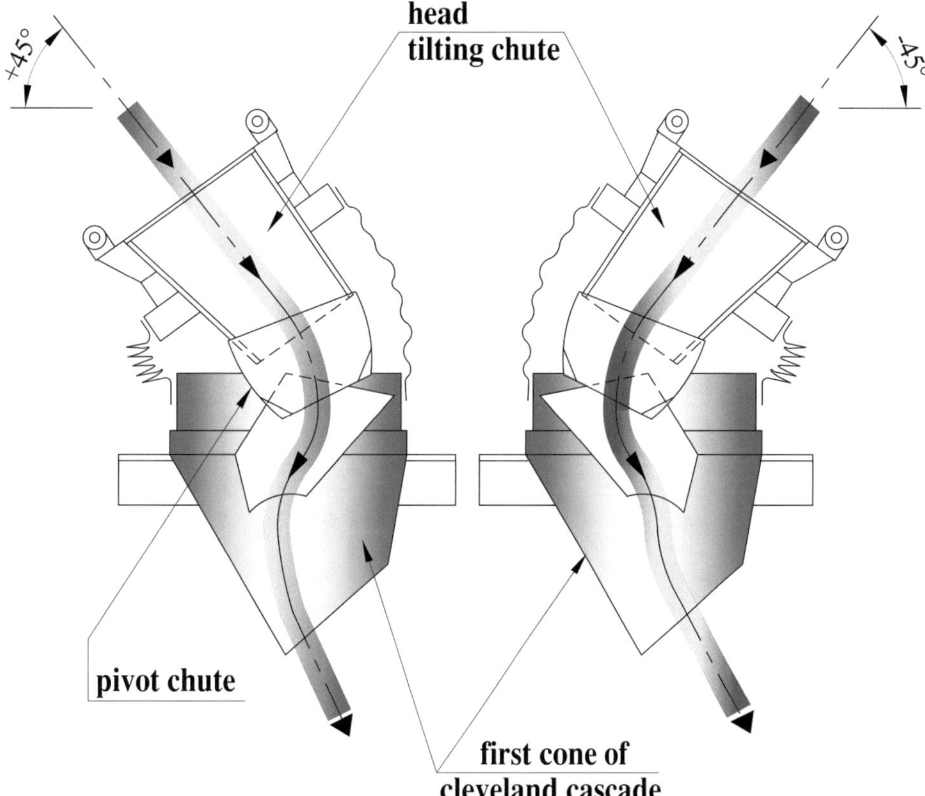

Figure 277. The pivot chute with which the head chute of boom conveyor can be tilted forward (+45°) or backward (−45°), as the Cleveland Cascade chute is always upright (see Figure 275).

The distinctive characteristic of shuttle-type shiploaders is related to the geometry of the boom conveyor shuttle movement (Figure 271): the further forward the shuttle will be *extended* (to load the bigger vessels), the distance between fully *retracted* shuttle (and, hence, head pulley of boom conveyor) from the edge of the wharf will be greater. Thus, the loading of small vessels becomes a problematic, complex task required using of special "spacers" in order to take the vessel off the wharf, in seaward direction, to complete the loading of the vessel.

To solve this problem, the new Bedeschi shiploaders are equipped with a *tilting*, 8 m long head chute. This innovation, developed by the author, differs these shiploaders from others (Figure 278).

The dedusting telescopic Cleveland Cascade chute is hinged to the bottom of the *tilted* head chute via a special pivot chute (Figure 277), keeping Cleveland Cascade always upright. The head chute is provided with special baffles slowing down the material movement to the inlet of the Cleveland Cascade chute. The head chute, tilted (± 45°) by two hydraulic cylinders, increases the loading outreach of the fully extended shuttle by about +5.0 m, and also tilts the loading chute backward at -5.0 m, closer to the edge of the wharf to load small vessels or barges (Figure 276).

The construction of the pivot chute, providing spillage- and dust-free connection between head chute and Cleveland Cascade telescopic chute, and ensuring the vertical position of the Cleveland Cascade chute at any angle of the tilted head chute, is shown in Figure 277.

The Bedeschi shiploader is equipped with the hydraulically extended service platform which in use for maintenance, repair, and cleaning of the Cleveland Cascade telescopic loading chute.

The boom of the shiploader can be tilted to the rest position (+65°) in about 5 to 6 min to be automatically locked to the A-frame (Figure 273).

Following is a comparison between the shiploader with a shuttle moving *inside of the boom structure* and the shiploader with a shuttle extended *from the boom structure*:

1. Stress on the boom structure

   Shiploader with shuttle conveyor moving inside of the boom

   - The 20-tons shuttle carrying a telescopic loading chute (weight of an extended clogged chute can reach 20 tons), moves inside of the enclosed boom without local overstress on the boom structure.

   Shiploader with shuttle conveyor extended from the boom structure

   - The heavy shuttle extends of from the boom structure and creates a large bending torque and local overstress of the boom structure, especially on the outer section of the boom.

2. Environmental considerations.

   Shiploader with shuttle conveyor moving inside of the boom

   - The boom is fully enclosed (except the narrow groove in the bottom) regardless of the shuttle position.
   - No dust emission from the internal boom conveyor.
   - The telescopic dedusting system reduces dust emission from the movable head hood of the boom conveyor.
   - Strong winds or drizzling don't interfere the shiploading.

   Shiploader with the shuttle conveyor extended from the boom structure.

   - The extended shuttle conveyor should be open from above to be constantly fed during its movement back and forth.
   - Fugitive dust generated by winds will cause unavoidable ecological problems.
   - The extended uncovered belt conveyor can't operate during strong winds or drizzling.

3. Maintenance and safety aspects

   Shiploader with shuttle conveyor moving inside of the boom

   - All maintenance operations are carried out inside of the enclosed boom structure.

Shiploader with the shuttle conveyor extended from the boom structure.

- Repair, replacement of broken parts and maintenance of the extended boom conveyor is troublesome operations from a safety point of view. The maintenance team should work outside of the boom using special safety equipment.

4. Capex considerations

   The cost of the shiploader with an extended shuttle conveyor is practically equal to the cost of a shiploader with an internal shuttle conveyor.

Note:

According to our experience (the ICL has been operating shiploaders with swivel boom, with non-swivel shuttle boom, with non-swivel shuttle boom and tilting head chute, with quadrant-type radial and with quadrant- type linear shiploaders), the *shiploader with non-swivel boom and internal shuttle conveyor* is the optimal solution for the most of conventional load-out wharves because it is the simple, reliable, low-maintenance and relatively inexpensive loader that comprises three movements: the travelling along wharf, the shuttling along the boom, and the tilting of the boom between +65° and -10° to travel from hold to hold (and to rest). The control of the loading from the deck of a vessel is also much simpler than the control of shiploaders with swivel booms (according to the personal experience).

Such a shiploader, especially *when equipped with a tilting head chute*, is capable to reach every spot of holds of big or small vessels, even if cranes or other obstacles are installed on the deck of vessels.

Other types of shiploaders.

The non-swivel-boom travelling shiploader towing attached inclined belt conveyor (stacker) that is fed by a "chain" of pivot-connected belt conveyors from a storage, is shown in Figure 278 and Figure 279 (Finland).

Advantage:

The wharf is free from the overhead conveyor bridge and from the floor-level fixed belt conveyors.

Disadvantage:

Each movement of the shiploader along the wharf/vessel requires the labour – and time-consuming rearrangement of the "chain" of pivot-connected feeding conveyors.\

Figure 278. Travelling shiploader with the shuttle moved inside the boom and with slope conveyor fed by the "chain" of movable pivot- connected belt conveyors (Finland).

Figure 279. Pivot connection of belt conveyors feeding the travelling shiploader.

Figure 280. Shiploader with the shuttle conveyor extended out of the boom structure (TAKRAF Tenova).

Shiploaders Fed by Stackers

There are piers or jetties that are not wide enough to install special bridges carrying feed conveyors. In such cases, it is accepted to use a special belt stacker connected to and towed by the shiploader and lifting the material from the floor-level fixed belt conveyor to a height sufficient to feed the boom conveyor (usually 20 m÷25 m).

This solution has following disadvantages:

- the stacker conveyor cannot be enclosed, so dust emission is a serious ecological problem.
- additional problems arise during rains and strong winds.
- open inclined stacker is difficult to inspect and to maintain.

But... sometimes, after profound analysis, such solution can be accepted as the optimal one.

Shiploaders fed by stackers are shown in Figure 281.

Figure 281. Travelling shiploaders fed by floor-level conveyors through stackers connected to and towed by the shiploaders (TAKRAF Tenova).

5.6 Quadrant-type Radial and Linear Shiploaders

Quadrant-type radial shiploaders are a special type of shiploaders for use in the exceptional cases (e.g., for offshore berths) (Figure 282).

The weight (and, accordingly, the cost) of the quadrant shiploader is about 2.5÷3 times higher than the weight/cost of a conventional travelling shiploader.

For example, the weight of a radial shiploader, loading ships up to 35,000 dwt at a capacity of 1,600 tph (port of Ashdod, Figure 282, Figure 283), is about 1,000 tons, as the weight of a traveling swivel boom shiploader, loading vessels of up to 65,000 dwt at a capacity of 1,600 tph and operated on the neighboring wharf, is about 350 tons.

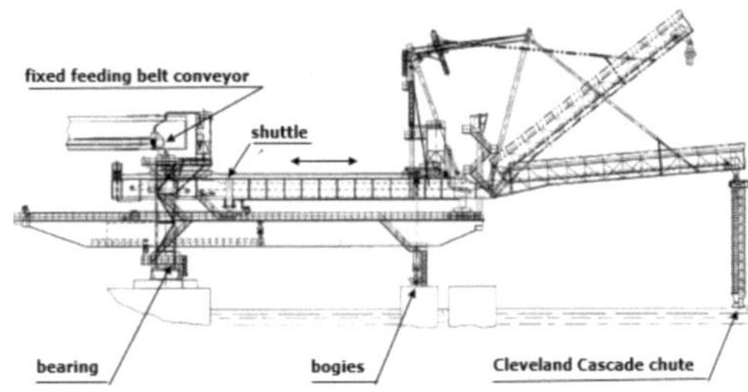

Figure 282. Quadrant-type radial shiploader.

Figure 283. Quadrant-type radial shiploader with Cleveland Cascade telescopic chute.

Radial shiploaders.

This type of shiploader consists of a bridge-type lower structure (portal) with a pivoted tail end. The seaward side of portal is supported by motorized bogies (Figure 282) that travel on circular rails.

The rails are maintained on the above-water structure supported by piles. The shiploader telescopic upper structure (with shuttle belt conveyor) travels forth and back along the bridge. The loading chute is suspended to the head chute of the belt conveyor. The combination of the bridge slew movement and the shuttle travelling on the upper structure (boom) allows the shiploader to load big vessels. To load all holds of big vessels, the extended upper structure of the quadrant shiploader must very long (Figure 284). Thus, the dimensions, the weight, and accordingly the cost of the shiploader are drastically increased. To overcome the problem of loading big vessels with no shifting, sometimes two or three smaller radial shiploaders can be used to load the same big vessel.

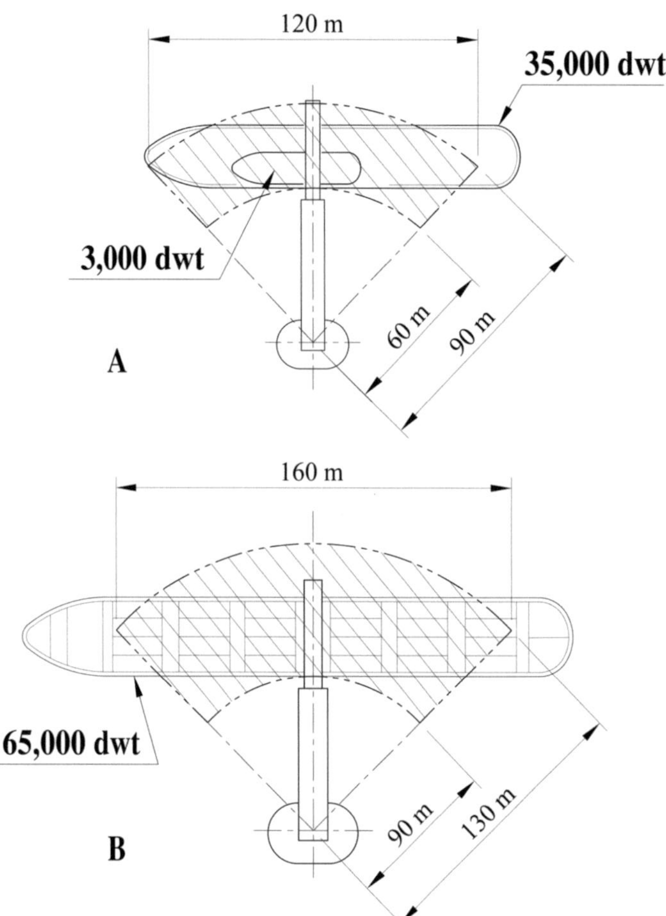

Figure 284. Radial shiploaders: ranges of shiploading areas for shiploaders designed for vessels up to 35,000 dwt (A) and designed for vessels up to 65,000 dwt (B).

Quadrant-type radial shiploaders are recommended only for special applications,

when shiploaders of other types are unable to perform the loading (e.g., when it is necessary to move the loading equipment offshore to deeper water).

Linear shiploaders

Quadrant-type *linear* travelling shiploaders (Figure 285, Figure 286) travel linearly along vessel and can reach every spot of holds using their linear and shuttle movements.

Advantages of this type of shiploader:

1. Allow to load the vessel at some distance from a wharf, in deep water.
2. Low height of tail section of the boom conveyor, so the low stacker is required for feeding the boom conveyor.

Disadvantages of this type of shiploader:

1. High Opex.
2. High Capex.

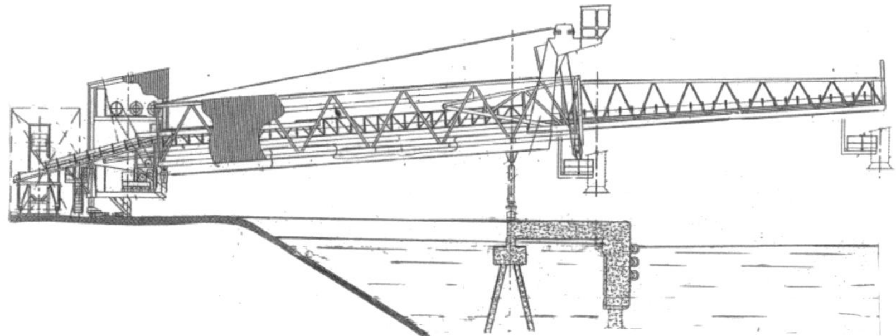

Figure 285. Quadrant-type linear travelling shiploader.

Figure 286. Quadrant-type linear travelling shiploader (USA).

5.7 Floating Shiploaders

No many floating shiploaders are in use today in the world, and most of them were manufactured by Bedeschi (Italy).

The floating shiploader consists of a fixed swivel shiploader (with hoppers and feeding conveyors) assembled on the deck of a pontoon, barge, or small carrier.

One of shiploaders of this kind could be seen in Figure 287, where shore belt conveyors or shore grab cranes feed two receiving hoppers, installed on the pontoon. The internal inclined belt conveyors are fed from the hoppers and transfer the bulk material to the fixed shiploader with the swivel and tilting boom. The shiploader loads a vessel, moored to the sea side of the pontoon.

This shiploader can be used also for offshore ship-to-ship loading operations.

Advantages of a floating shiploader:

1. The loader can be moored to any unequipped wharf with only one feeding belt conveyor or grab crane/s to start shiploading immediately.
2. The ship-to-ship transfer of bulk material can be carried out offshore.

Disadvantages of a floating shiploader:

All disadvantages of a fixed shiploader (see section 5.2).

Figure 287. Floating shiploader (Bedeschi, Italy).

5.8 Mobile Shiploaders

Mobile shiploaders (Samson, UK, and others) differ from conventional shiploaders in that they do not have a portal, an A-frame, bogies, so on. They are, basically, slope belt conveyors mounted on motorized rubber wheeled chassis, whose rear, loading section starts almost from the ground level, so they do not require the installation of rails on the wharf. Mobile shiploaders move or moved along the moored to the wharf vessels and can operate with a loading capacity of up to 2,000 tph (only in case the shiploader is fed with this capacity!).

The Samson-type shiploader can be fed from a bulk storage via a "chain" of movable and pivotally connected belt conveyors (Figure 288) or direct from a tipping truck/s discharger/s (Figure 289).

The sbelt conveyor can be tilted by hydraulic cylinders up to $+15^{\circ}\div+16^{\circ}$ and no more (a higher angle of belt conveyor can cause running back of most bulk materials, see section 2.1.14.4).

Mobile shiploaders with telescopic belt conveyors are manufactured by Superior Industries, USA, and other companies.

Figure 288. Samson loader for small vessels equipped with a short Cleveland Cascade telescopic chute (Finland).

Figure 289. Two truck dischargers feeding mobile Samson shiploader(UK).

Advantages of mobile shiploaders:

1. This is the right solution for the wharves that were not originally designed for heavy shiploaders traveling on rails.
2. They are the right answer to solve marketing problems requiring a sudden and fast shiploading solution.
3. They are the right solution for barge loading.

Disadvantages of mobile shiploaders:

1. Even to load small or medium-size vessels the shiploader must be long enough in order to reach a hold of the out-of-ballast vessel. For the big vessels, this problem is much more serious.
2. Any shiploader movement along a vessel when the shiploader is fed by a tipping truck discharger/s or by a "chain" of feeding pivoted belt conveyors, is a complex operational task.

   Any shiploader relocation is the time-consuming operation that significantly reduces average loading capacity and requires that the wharf be free from any other equipment.

   The average feeding capacity of the shiploader from tipping trucks (even using two tipping truck dischargers feeding one shiploader) is only 400 tph to 500 tph.
3. The cost of a large mobile shiploader is close to the cost of a travelling shiploader.

5.9 Shiploader Telescopic Dedusting Chutes

Dust emission is one of the major problems of shiploading operations.

The head pulley of a boom conveyor is usually located at a height of 20 m÷25 above the bottom of the hold of an empty vessel, so the free-fall speed of bulk material flow, especially on the first stage of loading, can reach 20 m/s to 25 m/s and the impact of the flow generates clouds of fugitive dust (Figure 290, Figure 291).

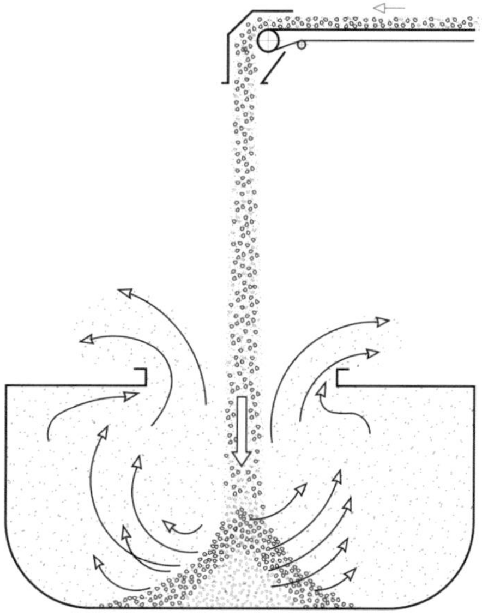

Figure 290. Dust emission during shiploading of dusty bulk material (without telescopic dedusting chute).

As an illustration, the dust emission during shiploading of fine phosphate with a non-effective dedusting chute is shown in Figure 291.

Tin order to overcome this environmental problem and to operate in accordance with the very strict requirements of the Environmental Protection Agencies (EPA), today all shiploaders should be equipped with telescopic dedusting chutes.

Figure 291. Shiploading of fine phosphate with a non-effective dedusting chute.

The choosing of the optimal, effective loading chute is one of the principle conditions for the successful operation of any shiploader. Below we consider four types of the loading chutes:

5.9.1 Midwest Chute

The Midwest chute was designed and built as a telescopic chain of vertical *concentric* cones wrapped in a flexible shroud. The lower section of the chute carries a number of small, built-in bag filters (Figure 292).

The dusty air, generated during the impact of falling material, is sucked in and cleaned by the dedusting devices. The cleaned air emitted outside, as the dust is poured out back.

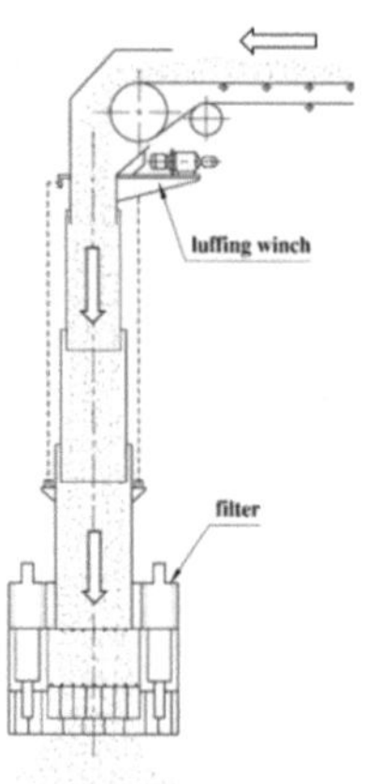

Figure 292. Principle scheme of a Midwest dedusting telescopic chute.

Advantages of the Midwest chute:

1. The chute is built on the principle of a telescope.
2. The shroud prevents side-windblown dust dispersion.
3. The dedusting system reduces dust emission after the material impacts the pile.

Disadvantages of the Midwest chute:

1. The principle disadvantage of such dedusting chutes is that the vertical *concentric* cones do not reduce the high speed of the material free fall, which leads to impact and dust generation. The generated dust, in turn, must be sucked in through bag filters. So, the *chute firstly "contributes" to the getting forth of dust and, secondly, cleans the dusty air and pours out the dust back.*
2. The bag filters are located in the centre of dust, so they require regular inspection, maintenance, and bag replacement.
3. During any malfunction of the bag filter system, the space between the cones and the shroud will be clogged with dust.

5.9.2 PEBCO Chute

The PEBCO chute also consists of vertical telescopic *concentric* cones wrapped in a flexible shroud.

The dedusting bag filter is installed on the top of the chute to suck in the dust from the free space between the cones and the shroud (Figure 293).

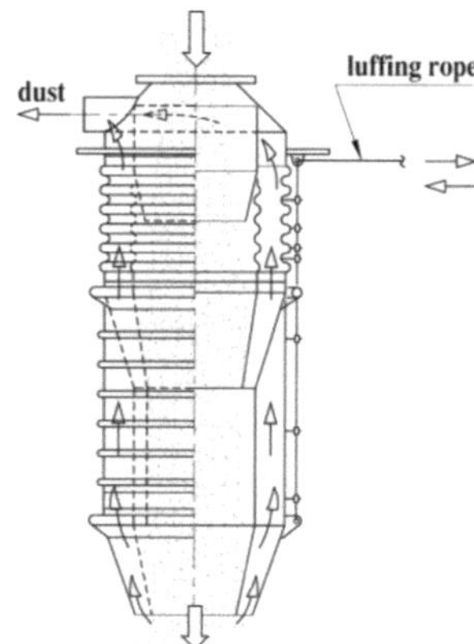

Figure 293. Principle scheme of the PEBCO telescopic dedusting chute.

Advantages:

1. 1. The chute is built on the principle of a telescope.
2. 2. The shroud prevents side-windblown dust dispersion.
3. 3. Dust emission is reduced.

Disadvantages:

1. 1. The principle disadvantage of such dedusting chutes (the same as the Midwest chute) is that the vertical *concentric* cones do not reduce the high speed of the material free fall, which leads to impact and dust generation. The generated dust, in turn, must be sucked in through bag filters. So, the *chute firstly "contributes" to the getting forth of dust and, secondly, cleans the dusty air and pours out the dust back.*
2. 2. If the bag filter is located on the top of the chute (to drop the dust back into the material flow), maintenance and bag replacement are difficult and unsafe operations. In the case when the bag filter is located at a considerable distance from the chute, the dust must be accumulated in big bags, and the replacement the filled bags with empty ones creates additional problems for the maintenance personnel.
3. During any malfunction of the bag filter system, the space between the cones and the shroud will be clogged with the dust.

5.9.3 DSH Chutes

DSH chute differs from all other loading chutes.

At first, the DSH chute is a fixed, non-telescopic loading chute.

This chute consists of an external open hopper that is suspended by means of springs to the upper fixed supporting structure, and an internal double-cone fixed (static) plug. The plug is located in the centre of the circle-form hopper outlet. Without bulk material in the hopper, the springs pull the hopper upward and the hopper outlet is blocked by the fixed plug (Figure 294).

When bulk material fills the hopper, and the *weight* of the material exceeds the springs' *tension*, the hopper is lowered and releases the hopper outlet. Now a bulk material is poured out through the open outlet (Figure 295). To keep the outlet open, the weight of the material in the hopper must be greater than the resistance force of the pulling upward springs.

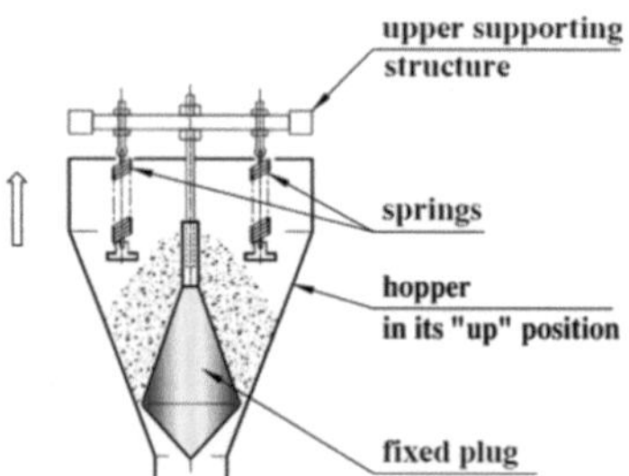

Figure 294. The DSH chute when springs pull the hopper upward and block the outlet.

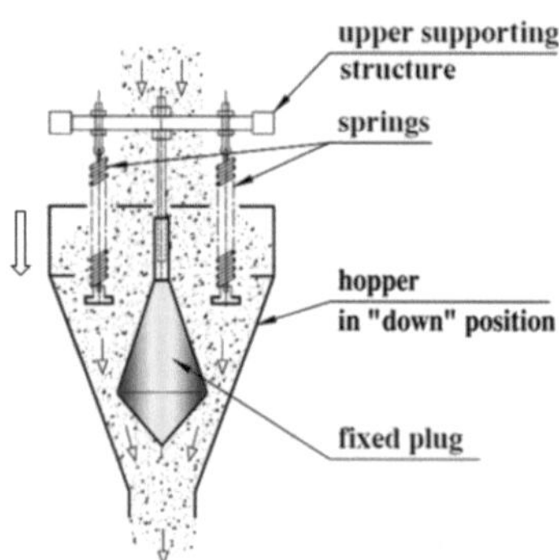

Figure 295. The DSH chute when weight of the material with the hopper exceeds the pulling of the springs; the hopper lowers and opens the outlet.

When the feeding capacity (and, hence, the weight of the material in the hopper) is smaller than pre-calculated capacity, the hopper outlet will be closed by the plug, and while the feeding capacity is larger than the maximum outlet capacity, the hopper will be overloaded.

So, uniform feeding and free-flowing homogeneous bulk material are two *must* conditions for the normal operation of this system.

Advantages:

1. During filling the hopper, the material flow is partially released from the excess air.
2. The initial falling velocity of the material (from the hopper outlet downward) is reduced.
3. Simplicity and low cost.
4. Can be recommended for factories operating in enclosed spaces and with small and constant loading capacities of homogeneous bulk materials.

Disadvantages:

1. The chute is *not* built on the principle of a telescope.
2. There is no shroud preventing windblown dust dispersion.
3. A non-uniform material flow causes periodic stops and restarts of the loading process accompanied by considerable dust emission.
4. Non-free-flowing or sticky bulk material can cause bridging and blockage of the system.
5. Large foreign objects or big lumps can get stuck between the hopper and the fixed plug, causing shutting down of the loading with following manual emptying and cleaning of the hopper.
6. The system is not fit for large loading capacities.

5.9.4 Cleveland Cascade Chute

The principle of Cleveland Cascade chute operation differs from that of other chutes.

The Cleveland Cascade chute is a telescopic "chain" of *alternate interlocked, "zigzag"-inclined* cones connected with flexible strips. The strips control the cone angle appropriate to the materials to be loaded (Figure 296). The cones are wrapped in flexible shroud.

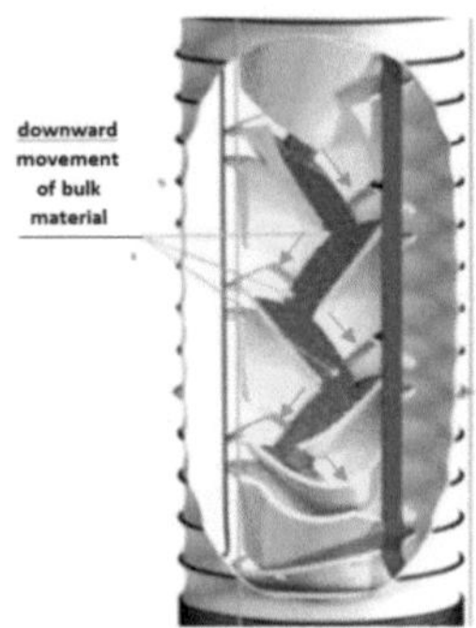

Figure 296. Principle of Cleveland Cascade chute operation.

The zigzag movement of a bulk material inside Cleveland Cascade chute is shown also in Figure 297.

At the very beginning of shiploading, the chute is retracted in order not interfere the installation of the boom above the hold of out-of-ballast vessel, and once the boom is above the open hold, the chute can be fully extracted. Four tilting probes control the height of the pile on which "rests" the chute skirt. A signal from the probes (Figure 297) automatically retracts the chute, lifting the skirt 0.2 m to 0.4 m at a time.

A recommendation for an operator: if the skirt "rest" on the top of the pile, the vessel will be loaded practically without dust emission (Figure 298, A).

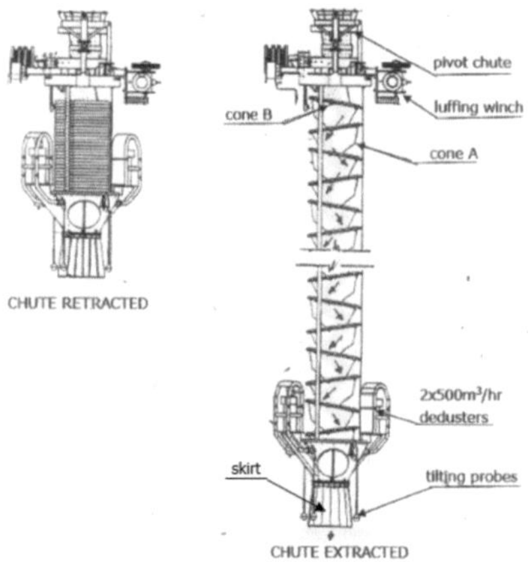

Figure 297. Cleveland Cascade telescopic chute with two additional dedusters.

Advantages of Cleveland Cascade chutes:

1. The chute is built on the principle of a telescope.
2. The shroud prevents windblown dust dispersion.
3. The zigzag movement of bulk material from cone A to cone B and so forth (Figure 296, Figure 297) leads to a significant decrease in bulk material speed. The terminal material speed is from 2.1 m/s to 2.5 m/s, in comparison with free-fall speed of material reaching is 20 m/s to 25 m/s. The low speed of material eliminates impact, effectively reduces dust emission, and lessens material degradation and segregation.
4. As an option, the chute can be equipped with two 500 Nm^3/hr insertable dedusters installed on the carrier of the chute (Figure 297).
5. The Cleveland Cascade chute is a simple, reliable, and low-maintenance dedusting equipment.
6. The Cleveland Cascade chute is the loading chute recommended by BAT.

Disadvantages:

In comparison with the loading chutes considered above, we did not find any significant drawbacks in the Cleveland Cascade chutes that we use for many years for different bulk materials and for different shiploaders.

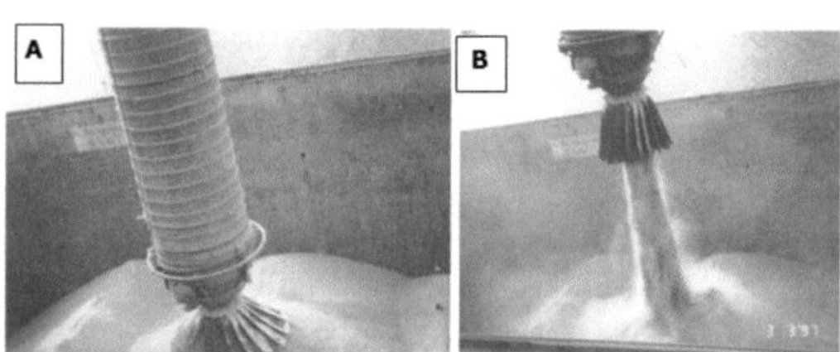

Figure 298. Shiploading of granular potash using a Cleveland Cascade chute as the skirt "rests" on the pile (A) and the skirt is lifted to a height of about 3 m above the pile.

Figure 298 (B) shows the same Cleveland chute loading the same bulk granular potash from a height of about 3.0 m from the top of the pile. The chute lifting caused considerable dusting. So, keeping the skirt of the chute as close as possible to the top of pile is the "*must*" condition for dustless shiploading.

New Development

In 2010, Flinders Ports (Australia) presented a new development in the shiploading: overturning and rapid emptying of the special 20-foot containers straight in the hold of the vessel. The main disadvantages of this loading are the discontinuous operation and the huge dust emission caused by the bulk material impact.

SHIP UNLOADERS

General

Ship unloading is a bulk material handling operation that can be carried out by various types of discharge machines. The choice of the optimal unloader depends on the type of wharf, on required discharge capacity, on characteristics of the bulk materials, on type of the vessel, on ecological and ATEX rules and regulations, and so on.

The most commonly used schemes of vessel discharge are:

- Vessel → ship unloader → hopper → truck → customer.
- Vessel → ship unloader → hopper → belt conveyor(s) → storage.
- Vessel → ship unloader → vessel.
- Vessel → ship unloader → hopper → truck → storage.
- Vessel → ship unloader → belt conveyor → storage/truck.

The results of systematization of various types of ship unloaders are presented below:

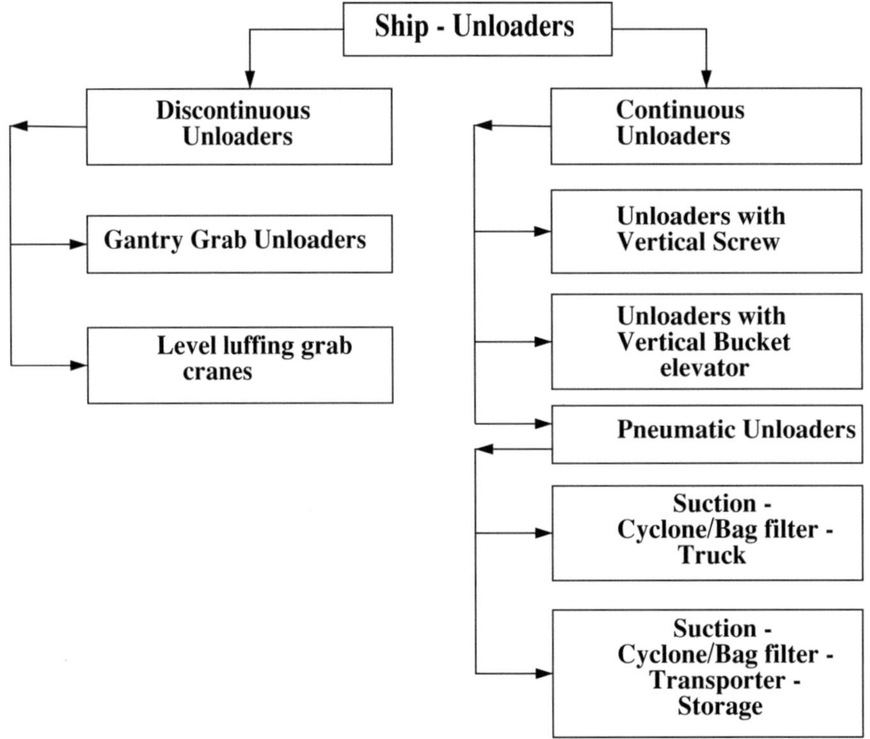

5.10 Gantry Grab Unloaders

Today a gantry grab unloader is the most widely used port equipment for discharge vessels carrying such bulk cargo as coal, sand, gravel, bulk fertilizers and so on.

The principle scheme of the unloader is shown in Figure 299.

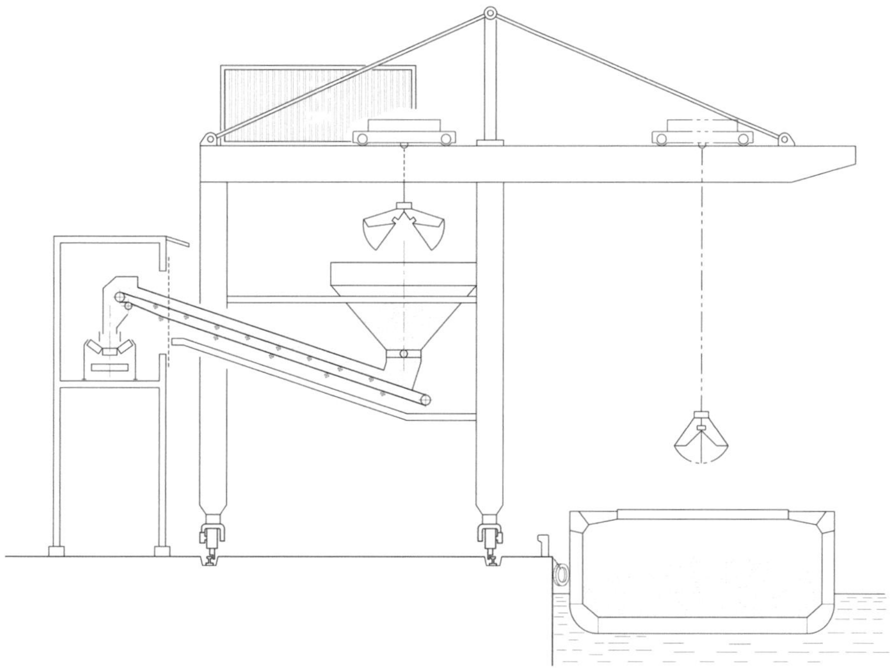

Figure 299. Principle scheme of gantry grab unloader.

Figure 300. Typical gantry grab crane discharge operation (TAKRAF Tenova).

A typical travelling gantry grab unloader moves on rails installed along the wharf and consists a of portal mounted on motorized bogies; an A-frame; a boom with a mobile grab carrier; an operating cabin and a receiving hopper mounted on the crane portal. The grab carrier moves along the boom.

The empty grab is lowered into a hold, opens its clamshells (Figure 301), and is filled with the bulk material during closing of the clamshells. When the grab is lifted, taken out of the vessel (Figure 302) and the carrier stops above the hopper. The clamshells of the grab are opened and the material is poured into the hopper, or into the special hopper mounted on the wharf for loading trucks or for loading a takeaway ground-level belt conveyor.

Figure 301. The grab in the hold of a vessel.

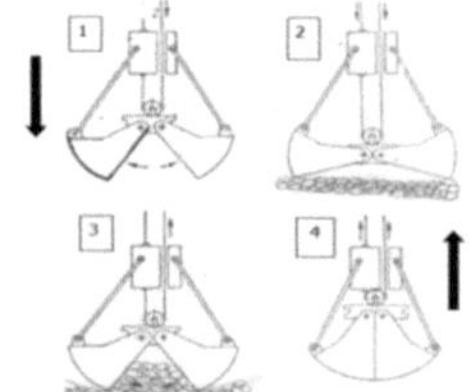

Figure 302. The principle schemes of grab operations.

To release the edge line of the wharf during non-operational periods, the outer section of the boom can be tilted upwards or to be swivelled along the wharf.

The use of a number of the unloaders, concurrently discharging one vessel onto a wharf belt conveyor or into receiving hoppers installed on a wharf and loaded trucks, can ensure required discharge capacity.

5.11 Level Luffing Grab Cranes

Level luffing grab cranes (Figure 303, Figure 304) automatically keep the grab at a constant height while the boom is raised or lowered. These cranes are manufactured in different versions. Figure 304 shows a portal crane with a swivel A-frame, and Fig. 305 shows a tower crane with a slewing upper structure.

At the beginning of unloading, when the hold is filled with the bulk material, the grab can be easily filled within 15 to 25 seconds (the full discharge cycle usually takes from 1.5 min to 2.5 min), and this unloading capacity is the maximum (so-called "cream") capacity. The only problem at this stage is the overfilling of the grab when crane operator should open the overfilled grab, discharge the excess material, and closes the grab again, so this operation requires more time and causes a reduction of the average discharge rate.

At the final stage of unloading, remains of bulk material should be collected in heaps by a front-end loader to facilitate filling of grabs.

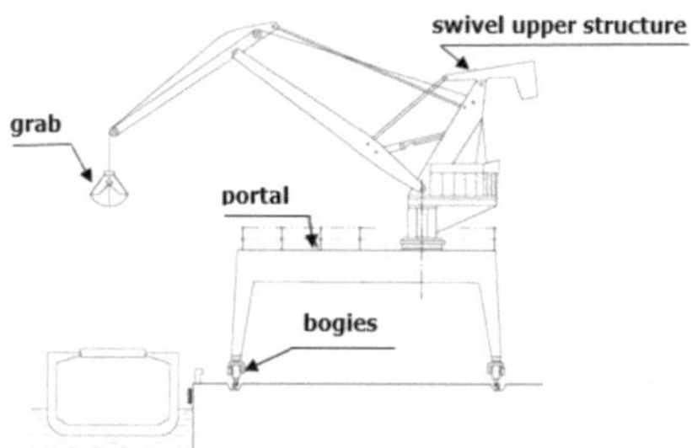

Figure 303. Typical luffing grab crane.

Figure 304. Ship unloading by luffing grab crane (Verstegen Grabs, Netherlands).

The level luffing grab crane can reach a discharge capacity of 800 tph to 1,000 tph. For example, a grab crane with a lifting capacity of 35 tons (19 m^3), a grab weight of 12 tons at self-weight of the crane about 100 tons, has an average discharge rate of about 1,000 tph. The common grab lifting capacity of such cranes is between 10 tons and 25 tons of bulk material.

Advantages of grab cranes:

1. Versatility: they can be used for unloading and for loading of various bulk materials as well as for operations with general cargo.
2. High discharge capacity that can reach 1,000 tph or more, at a low power consumption of 0.2 kWh/t ÷ 0.4 kWh/t.
3. A number of cranes, installed in row to discharge one vessel, provides any required capacity.

Disadvantages of grab cranes:

1. Spillage during movement of the loaded (especially overloaded) grab from the vessel to the hopper, causing contamination of a sea and a wharf.
2. Dust emission during unloading of dusty bulk materials into the hopper.

Figure 305. The ship-unloading using grab cranes, loading wharf hoppers. The trucks are loaded through the hoppers (Verstegen Grabs, Netherland).

To overcome the leakage problem, grabs with special sealing are manufactured by Verstegen Grabs, Netherland, and others.

According to the manufacturers, the new grabs prevent material losses and environmental pollution during the grab movement from vessel to hopper, but the dust generated during discharge of the grab into a hopper remains one of the main environmental problems of unloading vessels with the help of grab cranes.

Eco Hoppers are trying to reduce the dust emission during a grab discharge, by the use of built-in insertable dedusting systems.

5.12 Ship Unloaders with Vertical Screw Conveyors

Today the most of ship unloaders with vertical screw conveyors are manufactured by Siwertell–Cargotec Co. (Figure 306).

The principle feature of these unloaders is the exceptionally high rotational speed of the vertical screws conveyor reaching 500 rpm (for comparison, the rotational speed of industrial screw conveyors is between 40 rpm and 100 rpm). This high rotational speed is required to create significant centrifugal force pushing the bulk material against the housing to *increase friction between the bulk material and the housing*. The friction prevents the material *from rotating together* with the spiral and promotes the *movement of the mass of non-rotating material, by the rotating spiral, upward to the discharge chute* (see section 2.3).

Figure 306. Siwertell ship unloader (A) and typical pickup inlet device (B).

The know-how technology helps the Siwertell to produce unloaders that can reach up to a 3,000 tph unloading capacity ("cream" capacity) when discharging large vessels.

The specially designed spout or inlet feeder (Figure 306 B) picks up the material from the hold and feeds the vertical screw that lifts the material to the discharge point, where the material is transferred to the boom horizontal screw (or belt) conveyor/s and further discharged into a hopper or onto takeaway conveyor.

The high rotational speed causes intensive wear of the screws and the tubes (housings). The frequent bearings replacement is also the problem for the maintenance personnel.

The service life of the screws and tubes of Siwertell coal unloaders was about 3,500 to 4,000 work hours in 2010. The service life was supposed to be 8,000 work hours in unloaders manufactured in 2016.

The Siwertell unloaders are the unloaders that have special permit from ATEX for discharging such hazardous, explosive bulk materials as bulk sulphate and so on. Such unloaders are equipped with water-spray dust-suppressing systems and special safety doors that are opened in the case of dust explosion.

Advantages of screw unloaders:

1. Enclosed and dustless transporting.
2. Compliance with ATEX requirements.

Disadvantages of screw unloaders:

1. Frequent replacement of bearings, screws, and screw housings.
2. High power consumption (about 0.55 kWh/t).
3. High Opex and high Capex.

5.13 Ship Unloaders with Vertical Bucket Elevators

Unloaders with vertical bucket elevators are used for high-capacity discharge of such bulk materials as coal and ore (Figure 307).

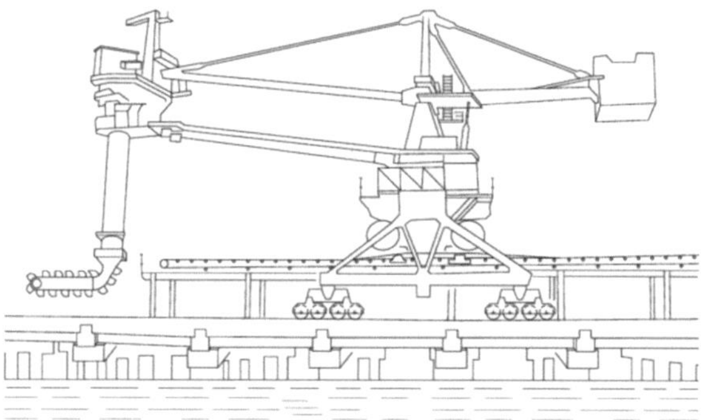

Figure 307. Typical ship unloader with bucket elevator.

The material is picked up by the short, horizontal inlet section of the elevator, lifted by the vertical section, and transferred to the boom belt conveyor loading a wharf/jetty takeaway belt conveyor.

Two bucket unloaders can discharge one vessel with the double capacity and transfer the material to the same takeaway belt conveyor.

Advantages of bucket unloaders:

1. High and constant discharge "cream" capacity from 1,000 tph to 3,000 tph.
2. Capability to discharge consolidated and agglomerated bulk materials.
3. Minor dust emission during unloading.

Disadvantages of bucket unloaders:

1. High Capex because of the cost of the high-capacity chain bucket elevator. The portal and upper structure must be strong enough to withstand large overturning moment, so the structure of the unloader is massive and heavy and, hence, expensive.
2. Because of above noted disadvantage no. 1, the load on the wharf is much higher than the load from other types of ship unloaders.
3. At the final stage of unloading, the remaining bulk material should be collected in heaps by shovel to facilitate the ending of the unloading.
4. High power consumption (kW/ton).
5. Bucket elevator, operated inside of the hold, requires regular inspection, cleaning and maintenance, resulting in relatively high Opex.

5.14 Pneumatic Ship Unloaders

Pneumatic ship unloaders use a vacuum pneumatic conveying systems (see section 2.9) to discharge such free-flowing bulk materials as dry powders, grain, light polymeric granules and so on, from bulk carriers and barges (Figure 308).

The principle scheme of a pneumatic vacuum unloader is shown in Figure 309.

The swivel booms of the unloaders carry vacuum telescopic pipe moved along the boom. The vertical section of the pipe with a rotating suction nozzle is lowered into a hold and, once in contact with bulk material, sucks in the bulk material from a pile by vacuum airstream.

The air-and-material mixture is directed by the pipeline to a receiver /bag filter, where the bulk material is separated from the air and transferred (via rotary valve) by screw conveyor or by belt conveyor straight to the storage, or to be loaded into trucks.

Figure 308 shows a two-boom fixed, swivel and telescopic grain unloader transferring the air–grain mixture straight into cyclones, that load belt elevator/s. The elevator feeds an overhead conveyor with tripper that successively fills the row of silos.

Figure 308. Two-boom fixed swivel and shuttle pneumatic grain ship unloader.

Figure 309. Mobile pneumatic ship unloader (Neuero).

A conventional pneumatic ship unloader consists of a centrifugal (or Roots) fan, vacuum pipes, and two bag filters (operating successively) with horizontal and inclined screw conveyors, loading trucks or takeaway belt conveyors.

A pneumatic ship unloader, shown in Figure 309, discharges a powder from a vessel through vacuum pipeline straight into the baghouse. Screw conveyor/s, mounted at the bottom of the baghouse, transfers the material to the chute/s loading trucks.

Kovako pneumatic discharge systems differ from conventional pneumatic unloaders. Such systems use deep vacuum systems to suck in fine bulk material (cement, phosphate, flour, and so on) from a vessel at high solids-to-air ratio (the pipelines are relatively short) and to convey it into baghouses, where the material separates from air. The screw conveyors, mounted on outlet of the baghouse hoppers, successively fill one of two positive pressure dense phase pneumatic transporters, conveying the bulk material to outer storage using compressed air supplied by the special compressor.

The floating trans-shipment terminal includes two dense phase pneumatic systems (Kovako) with an average unloading capacity of 300 tph of cement, alumina, or similar powders.

Advantages of pneumatic ship unloaders:

1. Clean, dustless operation.
2. No dusting during picking up the bulk material from the hold of the vessel.

Disadvantages of pneumatic ship unloaders:

1. The highest specific power consumption among all other types of bulk unloaders (up to 1.0 kWh/t and more, see Table 2, section 2.1.1).
2. The stringent requirements imposed on the quality of the bulk materials. Lumps or large foreign objects can cause plugging of the pipe and stoppage of transportation until the manual pipe-cleaning is completed.
3. Limited discharge rates, as the increase of the rate requires huge air (and, hence, power) consumption and use of large dedusting systems at the outlet of the transport pipe.
4. The longer the conveying pipeline, the lower capacity of the system and the lower the solids-to-air ratio μ (efficiency) of pneumatic conveying.
5. The longer conveying distance, the higher terminal air velocity and the more the bulk material particles will be grinded by attrition.
6. Only limited types of bulk materials (cement, grain, phosphate and similar bulks) that are dry, free-flowing and homogeneous, can be recommended for conveying by pneumatic transport.

CHAPTER 6

Dedusting Systems in Bulk Material Handling

General

As noted above, the suppression or significant reduction of dust emission during conveying, transferring, loading, unloading, and storing of bulk materials is one of the most important tasks in the bulk material handling.

The Guidance does not address the use of dust-suppressing installations of moisturizing bulk materials to reduce dust emission (chemical fumes, sprays of water, and so on). For such bulk commodities as fertilizers, cement, grain and so on, the moistening methods are not accepted.

Further will be presented various dry dedusting systems widely used in the bulk material handling.

6.1 Gravity Separators

Gravity separators (or baffle chambers) are simple primary systems used as intermediate separating equipment located between the dust-laden airstream and the secondary dedusting systems (bag filters). The dust-laden airflow enters the separator, where it impacts the deflection plate and, as a result, the airflow velocity is sharply reduced. The coarse particles are settled out under action of gravity, gathered at the lower section of the silo and discharged through rotary valve (Figure 310).

The dusty air, containing most of fine particles, continues to move to the secondary dedusting system (baghouse dust collector), where the final air-cleaning takes place. The efficiency of a gravity separator can be improved if an insertable deduster will be installed on the top of the separator as it is shown in Figure 310.

In principle, the separator segregates and removes big particles from fine particles in the particles-laden air flow to reduce the load on the bag filters installed after the separator.

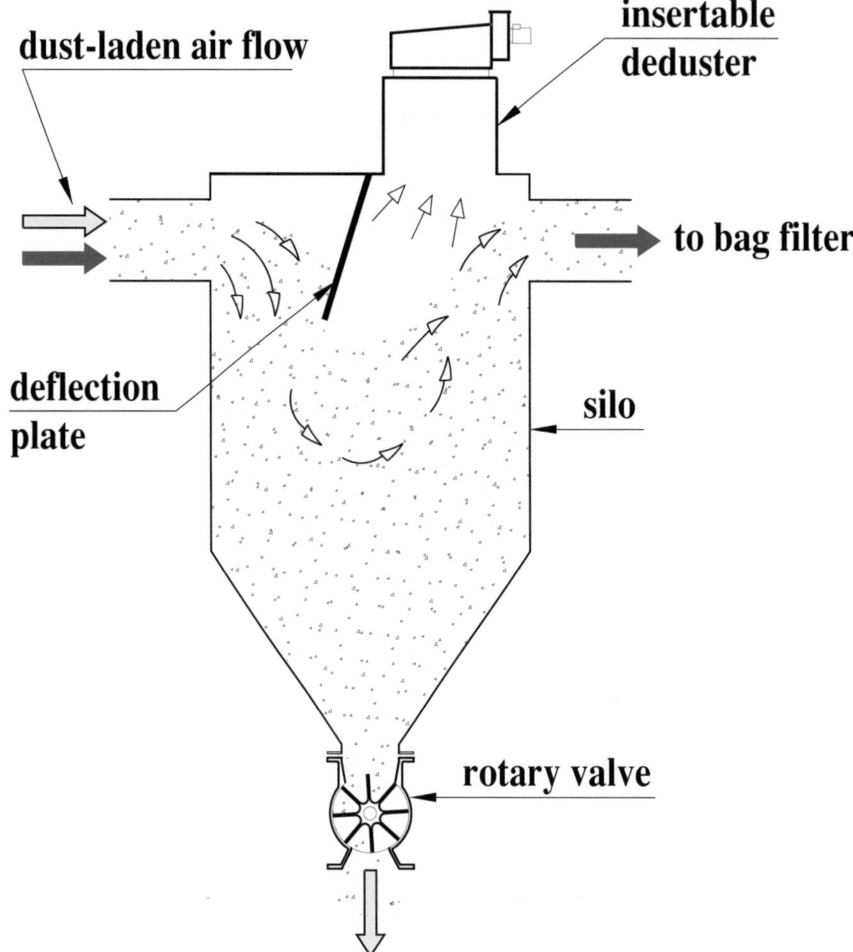

Figure 310. Work principle of gravity separator.

Advantages of gravity separators:

1. This primary separating reduces the excessive load acting on the secondary dedusting systems and increases total efficiency of dedusting.
2. The coarse particles are separated from fine particles.

Disadvantages of gravity separators:

1. Additional equipment (silo, rotary valve, sensors to prevent overloading of the silo, etc.) increases the energy consumption and maintenance requirements.
2. Additional space is required.
3. It is necessary to prevent bridging and ratholing at the outlet of the silo.
4. The cleaning of the silo is required in the case of material replacement.
5. Additional Opex and Capex.

6.2 Cyclone Separators

Cyclones, also referred to as cyclone separators or cyclone collectors, are usually used as pre-cleaning dedusting equipment that is much more efficient than gravity separators. The principle of cyclone operation is the removal the coarser, inert particles from the dust-laden airflow by friction forces (plus impacts). The cyclone is designed in such way that the most of heavy, inert particles move within the cyclone along a circular path and, being in contact with the walls, are decelerated by friction and fall, while the most of fine, light particles circulate without contact with the walls and exit along with the dusty air through the central pipe (Figure 311). Figure 312 shows the principle scheme of a cyclone separator that helps to calculate/design new cyclones or to re-check the main parameters of existing cyclones. The accepted inlet dust-laden air flow velocity is 12 m/s to 16 m/s, and the recommended optimal ratio (V) of the suction capacity Q (Nm³/s) to the cyclone cross-sectional area F (m²) is:

$$V = Q/F = 4Q/[\pi(D_c)^2] = 2.5 \text{ m/s to } 2.7 \text{ m/s}$$

The efficiency of a cyclone depends on particle size distribution of the dust (dimensions, shape and specific density of the particles) and, as it was noted above, the coarse and heavy particles are better separated and removed by a cyclone than fine particles.

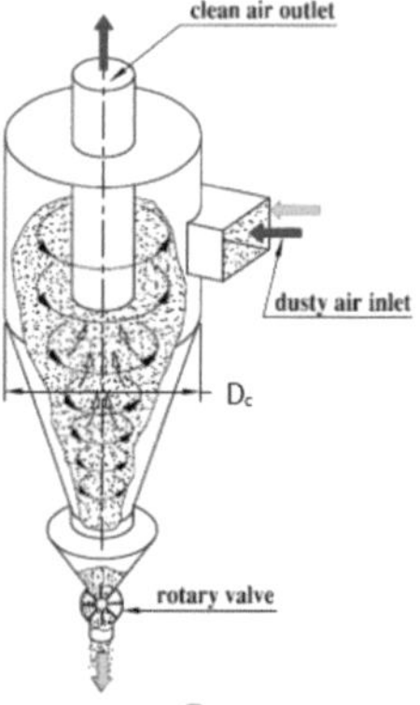

Figure 311. Work principle of a cyclone separator.

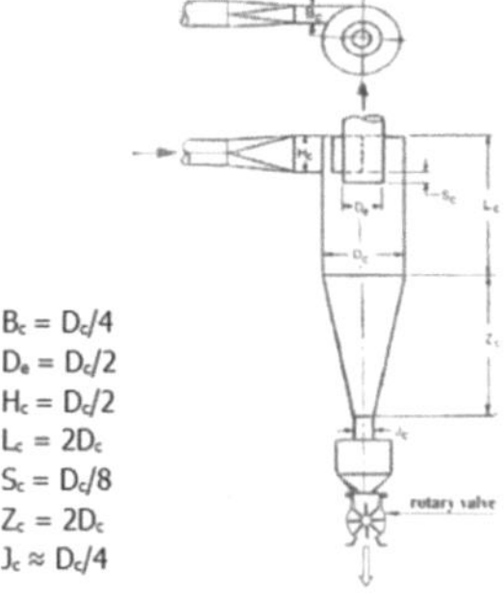

Figure 312. Cyclone separators (design recommendations).

Advantages of cyclones:

1. Simple and low-maintenance device.
2. Coarse particles (> 20 μm) are effectively separated and removed from the particle-laden air flow.
3. Low pressure drop (40 dPa÷80 dPa).
4. Cyclones reduce the load and increase efficiency of the secondary dedusting systems (baghouses).

Disadvantages of cyclones:

1. The collection efficiency of fine particles d ≤ 20 μm is only 92%÷95%, thus, the secondary dedusting system is required to remove the fine particles.
2. High humidity environment can cause sticking of fine particles to the inner surfaces of the cyclone.
3. Additional Opex and Capex.

6.3 Baghouses (Fabric Filters)

Baghouses are the most widely used dry dedusting systems in the world.

Baghouses are dedusting systems which can be divided into two main groups:

a. Air cleaning/dust collecting systems (baghouse dust filters) are systems that effectively clean dusty air from coarse and fine particles, gather the dust (particles) in the low, cone sections and transfer the dust through rotary valves and/or screw conveyors out of the system to fill big bags or to load takeaway belt conveyors, screw conveyors, and so on.

b. Air cleaning/dust disposal systems (insertable dedusters) are systems which effectively clean dusty air and then direct cleaned air out of the system. The dust is not gathered and poured back or onto the same belt conveyor or into the same silo from where dust was previously drawn.

A baghouse is a highly effective air pollution control system consisting of many filtering fabric cylindrical or envelope-type bags. The baghouse system using vacuum produced by the fan, to draw in heavily loaded dusty air with the initial concentration of 50 $g/m^3 \div 70$ g/m^3,.

When the dust-laden airflow enters the baghouse compartment(s), the part of large, heavy particles falls in the lower section of the hopper, and dust flow is filtered passing through filter bags as the fine particles stick on the outer surfaces of the bags, forming a layer or cake that improves the filtration efficiency and periodically removed by automatic bag cleaning system. The pressure fall in the bag filters is 80 dPa to 180 dPa.

The particle-laden air inlet velocity is usually between 2.5 m/s to 4.0 m/s. Figure 313 shows the principle scheme of a baghouse filter.

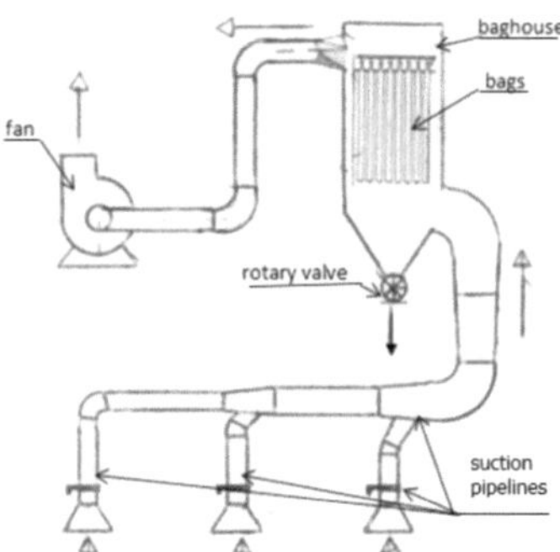

Figure 313. Principle of baghouse dust operation.

Most of baghouses with long (3m÷3.5m) tubular filters are equipped with reverse pulse jet cleaning systems, operating in such a way that one section of bags is cleaned by injection (via nozzle orifices) with short pulses of compressed air, as other sections continue the suction and the cleaning of dusty air. The cleaning principle consists in applying short pulses of compressed air. The air *pressure* and the *duration* of the air pulse determine the *speed* and the *dimensions* of the air "ball" of compressed air moving downwards along the tubular bag.

To produce such separated, high-speed "ball" of compressed air, (Figure 314), the pulse of compressed air must be very short (50 ms to 200 ms). The downward speed of the "ball" should be so high that the air *from the ball does not have time enough to be filtered out during the "ball's" movement.* The pulse of the compressed air must move downward as a solid ball!

The local "inflating – deflating" of a bag by the moving air "ball" results, as if "mechanical" bag-cleaning, and dislodging the layer of adhering to the outer surfaces particles.

The layers of ultrafine particles, constantly stuck to the outer walls of the bags, slightly increase the filtering resistance, but improve air cleaning efficiency.

The filtration speed recommended for the standard baghouses (*q*):

$$q = Q/F = (1.5 \div 3.5),\ [\text{m/min}],$$

where Q is the suction capacity (Nm^3/min) and F is the total filtration area (m^2).

The required area of filtration depends on the percentage of fine particles (≤10 µm) in the dust and on the initial load of dust (g/m^3):

- If PM_{10} < 80%, the required filtration area is about 30%÷40% of the filtration area required for the case of PM_{10} > 80%.
- If the load of dust (LOD) is ≤10 g/m^3, the required filtration area is about 50% of the filtration area required for LOD ≥ 50 g/m^3.

The height of the stack (Figure 315) is one of the important parameters of the dedusting system. The height can be determined using a special diagram presented TA Luft, 2002 (see below).

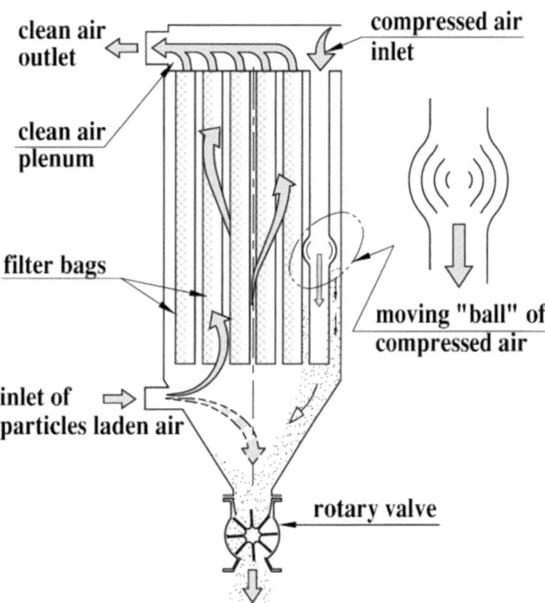

Figure 314. Work principle of reverse-air jet bags cleaning.

Figure 315. Operating industrial baghouse.

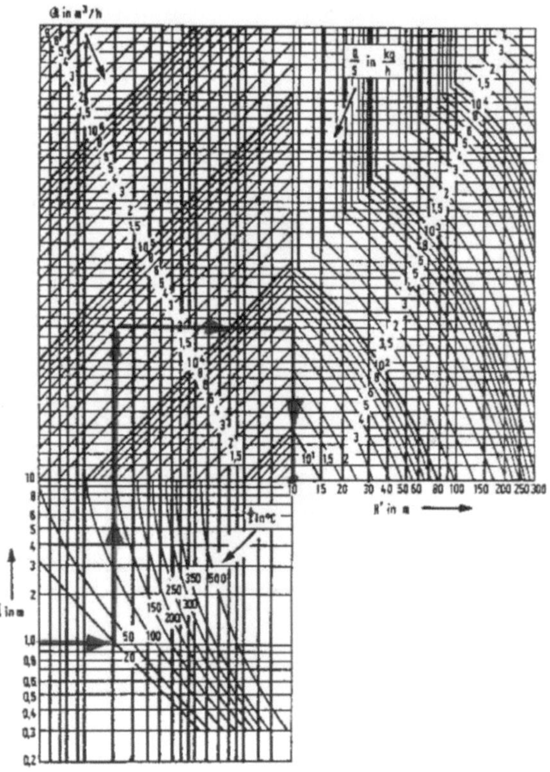

TA LUFT, 2002, Germany.

For example, for a stack diameter of 1.0 m, with a temperature +20°C, capacity of Q = 50,000 Nm³/hr and load of particles Q_m = 4.0 kg/hr at output emission of 5 µg/m³, the minimum height of the stack is about 10 m

Figure 316. Example of transferring dust gathered in a bag filter by screw conveyor back on belt conveyor.

Advantages of baghouse filters:

1. High effective cleaning of dry dust-laden air, up to 99.99%.
2. The dusty air can be cleaned to output emission of 5 mg/m^3.
3. The automatic bag cleaning (reverse pulse jet cleaning) allows the suction of dusty air through most of the bags while the part of the bags is jet-air-cleaned concurrently.
4. The bags made of special high-temperature fabric can withstand a temperature of +150°C to +160°C.
5. On the conveyor transfer points, the dust from bag filters can be sent back onto takeaway belt conveyor. Such an operation prevents the use of big-bags (and this is a labour-intensive and time-consuming operation) to remove the collected dust (Figure 316).

   Note:

   On transfer point the bag filters removes only the dust that arose as a result of a drop of bulk material from one conveyor to another and this is usually amounts to 5%÷7% of the total content of fine particles in the material. So, putting the collected dust back on the conveyor does not increase the *initial* percentage of fine particles in the bulk material.
6. Reliable, low-maintenance equipment.
7. Several connected baghouse compartments enable to clean to 200,000 Nm^3/hr and more of dust-laden air.

Disadvantages of baghouse filters:

1. Considerable footprint.
2. Periodically checking (monitoring) and replacement of damaged filter bags are required.

6.4 Insertable Dedusting Devices

Insertable dedusters (e.g., Dalamatic insertable dust collectors, DCE, UK; Martin Engineering, USA) are usually installed on transfer points of belt conveyors, on the skirt boards of takeaway belt conveyors at a distance of 1.8 m to 2.0 m from the feeding chutes. The deduster removes about 75%÷85% of the dusty air generated at the transfer point.

The recommended filtration air velocity, V_f, of a deduster:

$$V_f = Q/F = 0.032 \text{ m/s to } 0.035 \text{ m/s},$$

where Q is the capacity of dedusting (Nm^3/s) and F is the total area of bags (m^2).

Insertable dedusters can be used for dedusting conveyor transfer points, for dedusting of bins, bunkers, silos and so on.

Insertable dedusters can be equipped with silencers to reduce the operational noise to the permitted for industrial areas limit (70 dBa to 80 dBa, depending on the requirements).

An example of installed insertable deduster is shown in Figure 317, Figure 318.

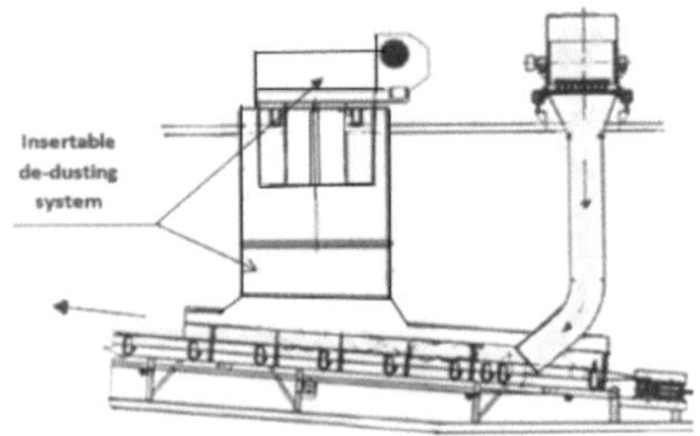

Figure 317. Example of installation of an insertable deduster.

The recommended inlet velocity of the dust-laden air at the inlet of the hood is 0.6 m/s to 0.8 m/s, and the normal pressure drop of the operating system is - 80 dPa ÷ -120 dPa.

Figure 318. Operating insertable deduster.

The Donaldson insertable dedusters are designed to suck in dust-laden air through a number of envelope-type, "flat" filters.

The cleaning of the flat filters is carried out in two stages: the compressed air supply (through blow pipe or jet tube) *inflates* the entire bag (first stage) an when supply of the air ceases, the air is filtrated from the bag outward, *deflating* the bag (second stage). Periodical inflation and deflation of entire bag dislodge the dust layer from the outer walls of the bag while the removed dust falls back onto belt conveyor or silo.

Advantages of insertable dedusters:

1. Effective local dedusting.
2. Compact and easy to maintain device.
3. Removed dust falls back onto a conveyor or into a silo, so no dust disposal problems.
4. Reliable and low-maintenance equipment.
5. Low Capex.

Disadvantages of insertable dedusters:

1. In contrast to the advantage no. 2: the dust is not collected and removed from the material but falls back into the material.
2. Relatively low operational vacuum, so the deduster should be installed very close to the source of the dusting.
3. Relative low suction capacities.

6.5 Wet Scrubbers

Wet scrubbers are effective air pollution control devices that are usually used for gas purification but also in bulk material handling facilities (in exceptional cases).

The principle of operation of Venturi-type scrubbers is that the contaminated air flow at high speeds of 60 m/s to 180 m/s is drawn through the Venturi throat, where water is radially injected. The water is atomized, creating a mist of tiny water droplets. Particles and water droplets are stuck together and transferred to the mist separator. From the separator, the clean air is directed to the atmosphere via exhaust fan, while the slurry/sludge is moved outside for further treatment.

The principle scheme of a typical ejection Venturi wet scrubber and the pressure distribution along the Venturi throat are shown in Figure 320 and Figure 321 [34, 35].

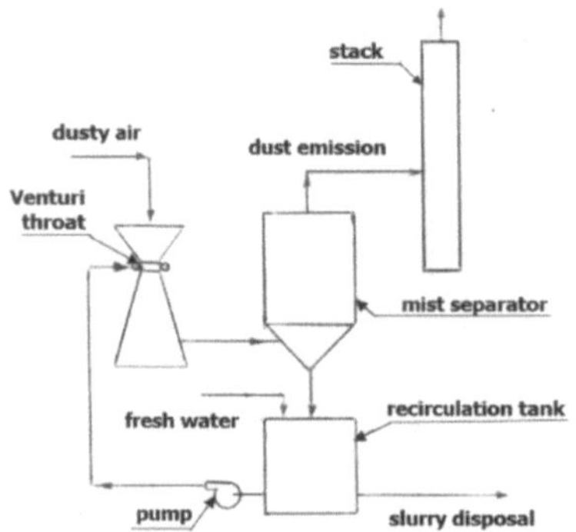

Figure 319. Principle scheme of Venturi-type wet scrubber.

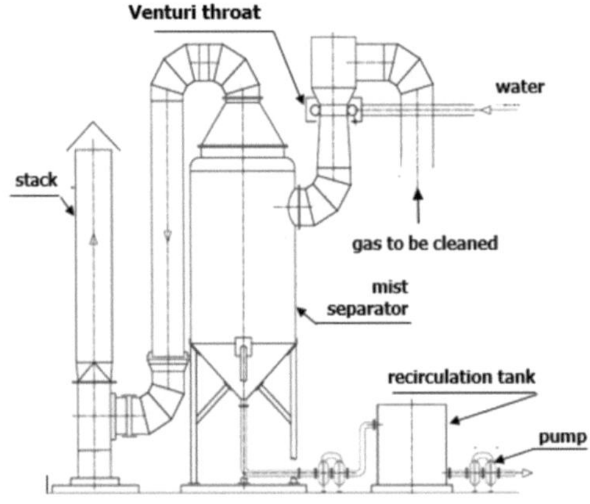

Figure 320. Industrial Venturi-type wet scrubber.

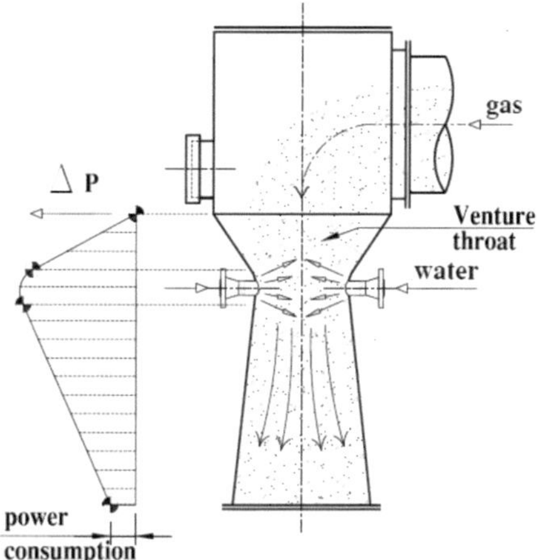

Figure 321. Air pressure distribution along Venturi throat of wet scrubber.

The technical parameters of a wet scrubber for dust removal are:

- Average cleaning efficiency: 70% ÷ 90% for fine particles, and 80% ÷ 99% for coarse particles.
- Scrubbing water pressure: 100 N/m^2 to 170 N/m^2.
- Airflow pressure drop: 50 N/m^2 to 200 N/m^2 (the efficiency of a Venturi scrubber depends on pressure fall: the higher the pressure fall, the higher the cleaning efficiency).
- Water-to-air ratio: Q_L/Q_G = (1.0 ÷ 3.0) L/m^3.
- Power consumption: 2.5 ÷ 6.5 kWh/1,000 m^3/hr. For example, the cost of the power consumption of Venturi wet 15,000 m^3/hr scrubber (fan and pumps) is about USD$5,000/year.

Counter-current flow scrubbers and cross-current flow scrubbers are also used for dust cleaning. Their characteristics are about the same as those of the ejection Venturi wet scrubbers, but Venturi scrubbers have the highest efficiency of dust cleaning.

Dry scrubbers are successfully used to remove acid gases (such as SO_2 or HCL) from combustion sources.

Advantages of wet scrubbers:

1. Can be used for cleaning of high-temperature dust-laden airflows.
2. Cab be used for cleaning wet dusty streams.
3. High particle loading does not affect the removal efficiency.
4. Loading fluctuations do not affect the removal efficiency.
5. ATEX approval for handling of flammable and explosive dusts.
6. Applicable for cleaning emissions with a high concentration of submicron PM (EPA, BREF).
7. Relatively low-maintenance requirements.

Disadvantages of wet scrubbers:

1. Abrasive and corrosive dust requires special resistant materials for the mist separator, cyclones, and piping (Fig. 320), resulting in high Capex.

   For example, several Venturi wet scrubbers were installed on potash rotary kilns to dedust the high-temperature (+180°C) outgoing dusty gas flow (28,000 Nm³/hr to 34,000 Nm³/hr per kiln). The flow is directed to the quencher (made from titanium, grade 7, to withstand corrosion), where injected water reduces the gas temperature to +60°C. The cooled flow enters the cyclone, where the particle concentration is reduced to about 1,500 μg/m³. Then flow is guided to the Venturi wet scrubber, where the concentration is reduced to the final concentration of 60 μg/m³÷ 120 μg/m³.

   The water can be circulated within the scrubber until the potash concentration in the water reaches 5.0 g/m³ (according to the maintenance team, the higher concentration causes clogging of the spray nozzles). The special sensors must monitor the concentration of the slurry to warn that it is necessary to replace it with fresh water.

2. Serious slurry-disposal problems.

   The waste water/slurry can be treated by being pumped by diaphragm pumps into settling ponds (insoluble dust!). Disadvantages of this method include the necessity of regular dewatering and drying of the ponds with further evacuation of the sludge, and this is complicated and expensive operation, requiring stop of the scrubber operation. In many cases the waste water is pumped back into the process. As an option, slurry can be discharged into the sewer net, but ... only after special permission that will be granted by the local authorities.

Fig. 322 Venturi wet scrubber for a potash dedusting system.

Fig. 322 shows the numerous patches covering the corrosion-worn carbon steel mist separator. This system stopped working many years ago.

As a combined solution, the soluble dry dust, collected into baghouse, can be dissolved in the water, using recirculation tank (Figure 320) or Venturi throat pipe (Figure 322.1), and the solution (after recycling) can be sent back into the production process.

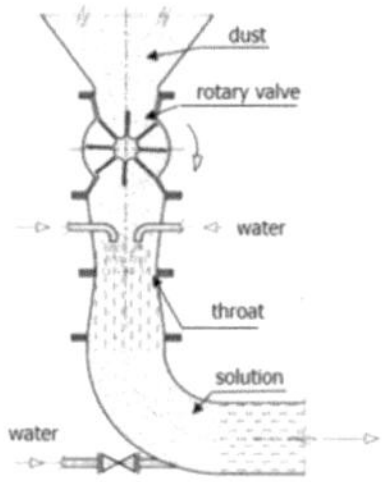

Figure 322.1. Soluble dust, collected in baghouse, is dissolved in water to be pumped in the production process.

Baghouse or Wet Scrubber—Which Is the Better Solution for PM Dedusting?

The correct answer depends on the characteristics and required capacity of the dust or gas that requires the cleaning.

It is recommended to use wet scrubbers to clean the contaminated gases:

- Explosive/flammable/hazardous dusts (e.g., dust of bulk sulphur).
- Contaminated high-temperature gases.
- Wet particulate gases.

Wet scrubbers are an irreplaceable solution for coal-fired plants (sulphur oxide emitting), for desulfurization of flue gases, and so on.

Note:

Our experience has shown the high efficiency of baghouses for the dedusting of such bulk materials as potash, phosphate, apatite in factories and ports. Baghouses represent a simple, reliable, low-maintenance, and cost-effective technical solution for dedusting, providing an outlet of air with a PM concentration of about 5 $\mu g/m^3$.

6.6 Electrostatic Precipitators (ESP)

The principle of electrostatic precipitator operation is that: high voltage (20 kV to 80 kV) is applied to the corona wires to form an electrical field between the wires and the grounded collecting plates (electrodes), and to ionize the air around the wires to create ions.

When dust-laden air flow moves, at a velocity of 1.5 m/s or below, between the plates and corona wires, the fine particles are charged by the ions (Figure 323).

Coulomb's law states that the electrostatic force between two charged objects (in our case, particle and plate) is directly proportional to the product of the quantities of charge and inversely proportional to the square of the separation distance between the two objects.

Coulomb's force causes the charged fine particles to be collected on the oppositely charged plates. In this way, the air is cleaned of particles and continues to move outward.

Removal of adhering particles from the collecting plates can usually be done without stopping the air-cleaning process.

There are three main methods of cleaning the plates from adhering particles:

- by rapping or vibration (fine particles, adhered to the plates, are removed by rapping or vibrating the plates either continuously or at predetermined intervals)
- by scraping with brushes
- by washing off with water.

The greater concentration of particles in the dust-laden air, the more time required for ion–particle interaction, that means the collecting plates must be longer.

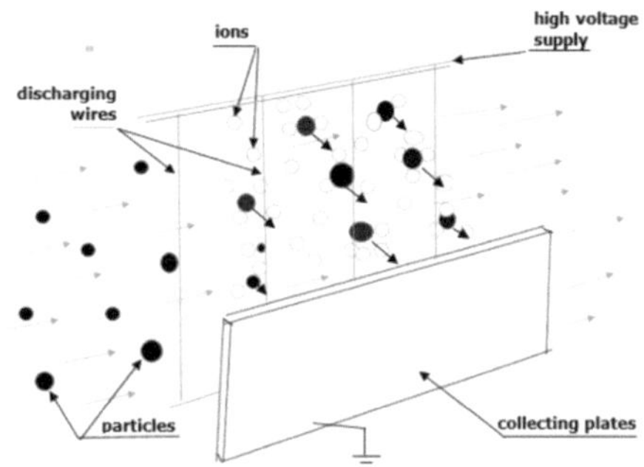

Figure 323. Principle of electrostatic precipitator operation.

The collecting efficiency of an electrostatic precipitator depends on the particle diameter (EPA – 600/8-77-020-a):

For particles with a diameter between 0.2 μm and 1.0 μm, the collecting efficiency is 93%÷ 96%, and for particles with a diameter between 4.0 μm and 6.0 μm, the efficiency is 99.7%÷ 99.8%.

Advantages of electrostatic precipitants:

1. Electrostatic precipitants are used only in exceptional cases when it is necessary to clean a huge amount (up to 400,000 m^3/hr) of flue gases, acid drops, or very fine particles (coal power stations, cement production facilities, etc.).
2. Dust-laden air is cleaned of ultrafine ($PM_{2.5}$) particles at a high collecting efficiency.
3. Air cleaning and cleaning of plates from adherent particles occurs at the same time.

Disadvantages of electrostatic precipitants:

1. Significant footprint.
2. Extremely high Opex.
3. Extremely high Capex.

CHAPTER 7
Samplers

7.1 Introduction

One of the problem with which the engineer often encounters is the choice of the sampler that would be the optimal solution for the given bulk material, for the specific technological conditions and for determined purpose (for analysis of chemical composition or particle-size distribution).

Today more than sixty companies located around the world are manufacturing various types of samplers. This variety proves that the universal, standardized sampler adaptable to all applications does not exist. Each sampler should be a correct solution for a specific problem.

The most important requirement to the sampler is that the extracted sample must be the representative one.

A representative sample is a small quantity of material the characteristics/parameters of which accurately represent the entire batch of material that should be in accordance with the specification of the bulk material.

The International Fertilizer Association (IFA, A/16/32, 2015) states that a number of sampling units for the shipment should be taken: the square root of the number of units (packaged product) or the square root of four times the total tonnage (bulk) present in the lot or shipment, with 10 units being the minimum. The best practice in the export ports is to take samples every 1,000 to 5,000 tons of loaded bulk materials.

7.2 Sampling of Bulk Solids

As noted above, the object of the sampling procedure is to get a representative sample that at the highest degree of probability represents characteristics of the transported bulk material [40, 41].

The representative sample should be extracted from the flow of the bulk material in accordance with the following basic principles:

a. The sample should be extracted precisely—the variance of the measured value must not exceed a given upper bound.

b. The sample should to be unbiased—meaning value without any systematic error.

c. The sample should be reproducible.

d. The sample-taking container/cutter/scoop should not be overflowed during cutting the material flow/layer and should be parked away from the material flow and from dust at the rest time.

7.2.1 Manual Sampling

There are two main methods of manual sampling:

a. Sampling from a stopped belt conveyor.

b. Sampling of a conveyed bulk material.

a. Manual Sampling from Stopped Belt Conveyor

The manual sampling operation shown in Figure 324 is the most reliable and representative sample-taking method.

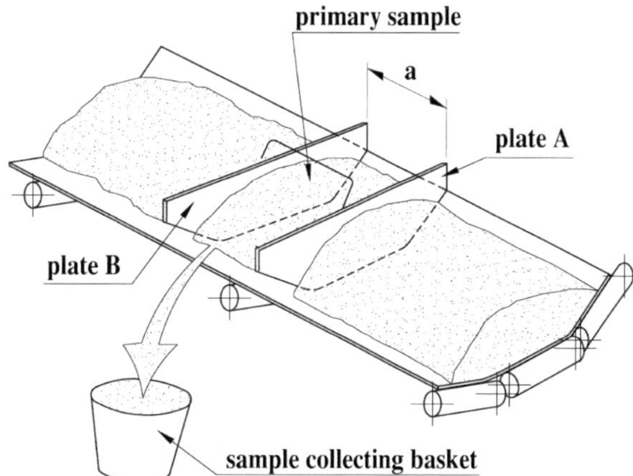

Figure 324. Principle scheme of manual sampling from a stopped belt conveyor.

The sample-taking procedure begins with the stopping of the loaded belt conveyor.

Next step: two plates (see Figure 324, A and B,) that are shaped to be fit with the profile of a loaded belt, are manually inserted into the material at an angle of 90° to the conveyor centre line, separating the "*a* " wide slice of bulk material, where *a* = 100 mm to 150 mm.

Next step: the manual brushing of the slice (separated by A and B plates) into the sample-collection container.

The weight of this primary sample is relatively high. For example, for a B1200 belt conveyor (speed V = 2.5 m/s, capacity Q = 1250 tph), the weight of 100 mm of a slice of bulk material taken from the conveyor belt (q_m) is:

$$q_m = 0.1 \times 1250/(3.6 \times 2.5) = 14 \text{ kg}.$$

The weight of the secondary representative sample which must be sent to the lab for analysis

is 100 g÷ 200 g, hence the primary sample should be *representatively* splitted/divided (manually or automatically).

Advantages of manual sampling:

1. The separated slice of the material, manually brushed out from the conveyor belt, includes all fine particles that were collected on the bottom of the layer of the material, on the belt (Fig.325). So, this sample represents a real cross section of the layer of bulk material on the belt.

   Note:

   Any polydispersive bulk material being transported on a conveyor belt is subject to the so-called "sifting" effect.

   The "sifting" effect is the process caused by vibration of the loaded conveyor belt on idlers: the belt slightly bounces, moving from one idler set to the next idler set (the belt deflection between two sets is 2%÷3% or 30 mm÷45 mm). As a result, the fine particles penetrate, slip between coarse particles and are accumulated on the bottom of the bulk material layer (Figure 325).

   This effect should be taken into consideration when one needs to take samples for particle-size distribution analysis. Fine particles are always accumulated on the bottom of the layer.

2. This method does not require Capex.

Disadvantages of manual sampling:

1. Undesirable periodical stops and restarts of the fully loaded belt conveyor.
2. Frequent stops/restarts reduce the average capacity of transporting.
3. The time-consuming and labour-consuming operations (relatively high Opex).

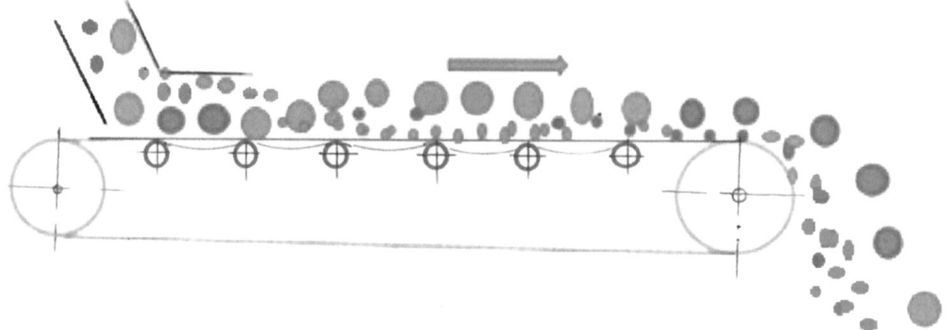

Figure 325. Principle scheme of the "sifting" effect.

b. Manual Sampling from the Flow of Bulk Material

Manual sampling of the flow of bulk material is carried out using a special sample cup that is inserted (through window in the conveyor head hood) into the flow by hand and cut the material flow (Fig. 326).

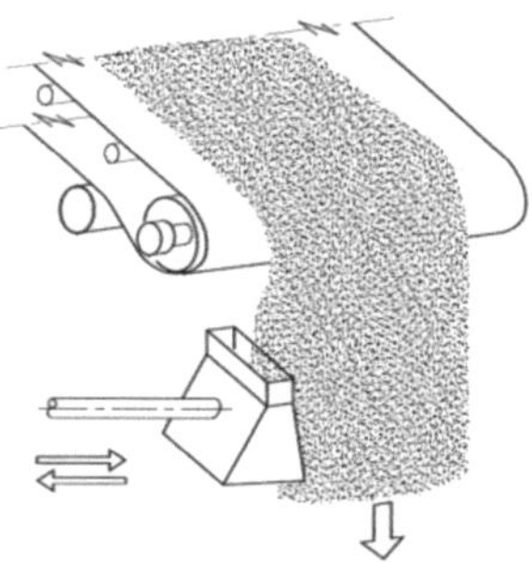

Fig. 326 Principle scheme of manual sampling.

The minimum time of this operation is about two seconds, and this time includes the movement in, cutting of the material stream and movement back. During this operation, it is necessary to fill the cup, but do no overflow - then the sample will not be a representative one.

For example, a cup, containing 1 kg of sampled material, will be filled in during two seconds with a material flow of about 1.8 tph capacity.

So, this method of sampling can be recommended only for small capacities of conveying.

Advantages:

1. Simplicity.
2. Recommended for belt conveyors of small capacities (< 2 tph).
3. No Capex required.

Disadvantages:

1. The holding and moving the cup by hand do not guarantee the right position or angle of the cup relative to the material flow.
2. It is very difficult to catch fine particles falling very close to the head pulley ("sifting" effect, Fig. 325).
3. The head hood must be equipped with a special open/closed "window" to allow the hand-and-cup movement back and forth.
4. Safety considerations.

7.2.2 Automatic Sampling

We are considering four types of automatic samplers:

7.2.2.1 Hammer-Arm Samplers

The hammer-arm sampler is the most widely used device for taking primary samples from moving belt conveyors.

The rotating hammer or rotating sample-taking scoop (or shovel) is attached to the rotating lever mounted above the belt conveyor. During the sample-taking operation, the bottomless scoop performs a rotating movement, accelerated by gravity, cutting the slice from the layer of moving bulk material, and pushing it forward into side-located sample-collecting container (Figure 327, Figure 328).

The higher the speed of the scoop movement, the better: the smaller amount of the bulk material will be taken as the representative primary sample.

After the primary sample is taken, the motor drive continues the scoop motion to its upper position, where it is stopped, ready for the next sample-taking operation.

To ensure the representativeness of the sample, the shape of the scoop trajectory should be adapted to the cross section of the loaded belt.

Brushes and rubber wipers must be properly connected to the scoop to ensure that fine particles, adhering to the belt, will be brushed by the scoop.

The coal samplers (Figure 329) usually take samples for chemical analysis only, so the representativeness of the primary sample (in terms of particle-size distribution) is not particularly important for the client.

The primary coal sample is transferred to a crusher. From the crusher, the sample directed into the splitter that cuts off a small but representative secondary sample to be taken to a lab for chemical analysis.

The rejected coal is collected by a special takeaway inclined belt conveyor, discharging the coal back on the main belt conveyor.

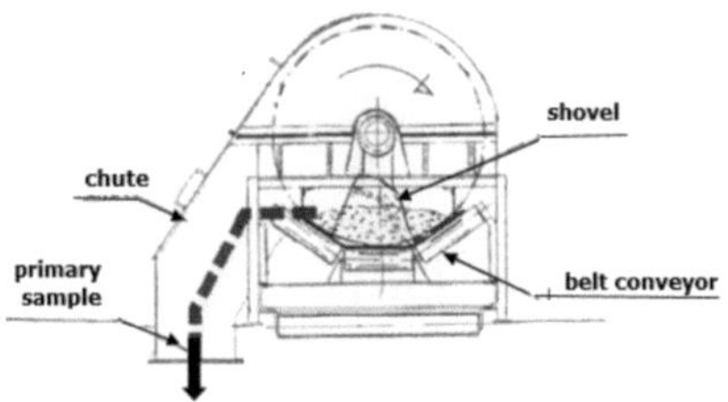

Figure 327. Principle scheme of a hammer-arm sampler operation.

Since the trajectory of the scoop movement is the circular, the special semi-circular bar bed (made from UHMW polyethylene) can be installed under the scoop instead of conventional sets of idlers (see Figure 57).

This arrangement helps to take a sample suitable for particle-size distribution analysis.

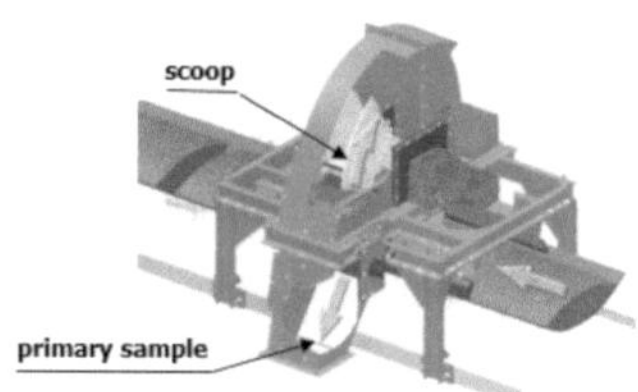

Figure 328. Hammer-Arm installation (Siebtecnnik GmbH).

Figure 329. Hammer-arm sampler installed on a coal belt conveyor (Siebtechnik GmbH).

Advantages of a hammer-arm sampler:

1. Simple, reliable, low-maintenance device widely used for automatic taking of primary samples from moving belt conveyors.

2. The sampler takes a relatively small slice of bulk material because of high-speed of scoop cutting the moving bulk material.

Disadvantages of a hammer-arm sampler:

1. In practice, the scoop does not scrub the belt, and some clearance (10 mm to 20 mm) is kept by the maintenance team to prevent accidental damage of the belt by the scoop.

   This clearance is not important for analysis of chemical composition of bulk materials, but remembering (see Figure 325) that fine particles accumulate on the bottom of the layer, such a sample (without fine particles) can represent a systematic error in measuring of particle-size distribution of the bulk material.

2. During the rest period (between sample-taking operations), the scoop and the sample-collecting container are open to dust entry, so regular cleaning of the sampler is required, especially when changing from one type of bulk material to another.

3. The primary sample, taken by hammer-arm sampler, is dropped by the scoop downward to be splitted further. To collect and direct the reject bulk material back on the same belt conveyor, an additional inclined belt conveyor (Figure 329) or special elevator are required.

7.2.2.2 Slotted Vessel Samplers

Slotted-vessel samplers are installed within the conveyor head hood (Figure 330) and used for receiving a sample by the horizontal movement of the vessel.

The vessel is moved from its initial waiting position through the flow of bulk material at a constant speed and with open bottom gate to remain empty up to the end position. At the end position, the bottom gate is automatically closed and the vessel moved back, across the flow, gathering a cross-sectional representative primary sample.

After reaching its initial, waiting position, the bottom gate of the vessel is opened (by means of a mechanical system of levers) and the vessel is emptied into a special sample-collecting container.

The obtaining a sample during the one-directional movement of the vessel prevents overflowing of the vessel.

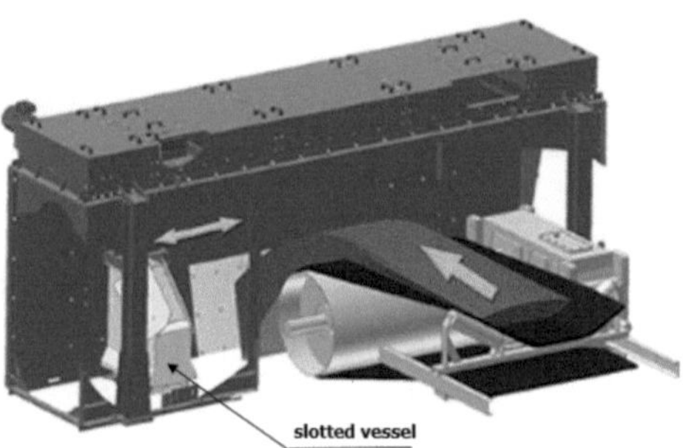

Figure 330. Principle of slotted vessel sampler operation (Siebtechnik GmbH).

The primary sample of the slotted sampler is directed into a primary splitter. The rejected material is removed by the takeaway belt conveyer as the sample is transferred to the secondary splitter.

The reject from the secondary splitter is gathered in a special bin.

Advantages of a slotted vessel sampler:

1. The slotted vessel moves at a constant speed and at angle of 90° through the material flow, taking a representative primary sample.
2. The sample-taking operation is fully automated.
3. The one-way sample-taking movement of the slotted vessel avoids overflowing of the slotted vessel.

Disadvantages of a slotted vessel sampler:

1. The slotted vessel, operated within belt conveyor's dusty head hood, is contaminated by the material, especially by wet, sticky dusty material, resulting in serious malfunctions.
2. Operation of the sampler within the enclosed head hood, where it cannot be visually inspected, makes the detection of mechanical failure is a difficult task.
3. Sloping belt conveyor or elevator is required to discharge the reject bulk material back on the main belt conveyor.
4. High Opex and high Capex.

Slotted-Vessel Piston Samplers

A cross-cut slotted-vessel sampler with a pneumatic cylinder pushes the rotating pelican-type sample cutter into the flow of material and pulling it back. Between the sample-taking operations, the cutter is out of the head hood of the belt conveyor.

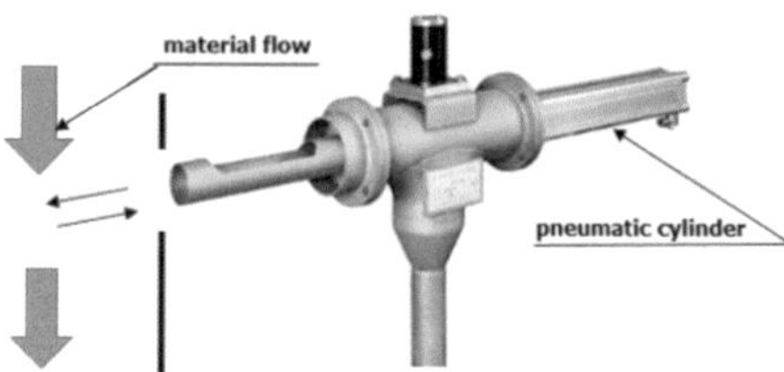

Figure 331. Slotted-vessel piston sampler (Siebtechnik GmbH).

The slotted-vessel piston sampler, shown in Figure 331.1, is a "self-made" sampler used in ports for taking primary samples from bulk material falling from head pulley of a conveyor. The relatively small volume of cutter ($V_{cut} \cong 2.5$ dm^3 or about 2.5 kg of sampled bulk material) and its back and forth movement causes the cutter to be even overflowed by the conveyor of very low-capacity. So, the device is recommended to take samples for chemical composition analysis only.

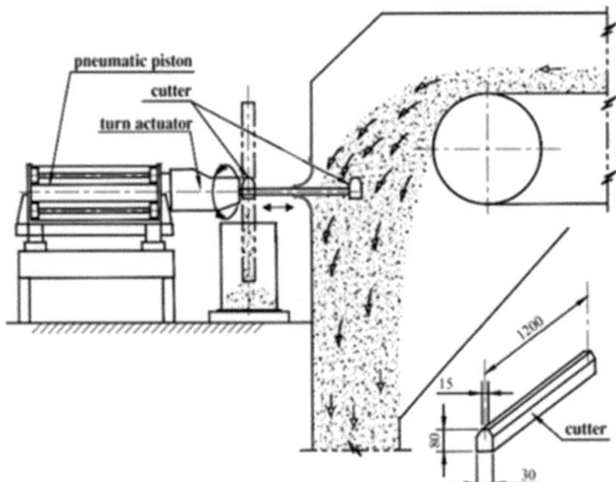

Figure 331.1. Slotted-vessel piston sampler for taking a primary sample for the analysis of chemical composition of a bulk material.

Advantages of the sampler:

1. The sampler is mounted out of the head hood, thus, between sample-taking operations it is not contaminated by dust.
2. Easy to inspect and clean.
3. Simple and inexpensive solution to obtain a sample of a bulk material for chemical composition analysis.

Disadvantages of the sampler:

1. The small volume of the open cutter and its back and forth movement cause overflowing of the cutter during the first seconds of cutting material flow, so this primary sample is not representative for particle-size distribution analysis, but it can be used for chemical composition analysis.
2. Applicable only for sampling of a low-capacity material flow.

7.2.2.3 Tilting-type Samplers

A tilting slotted-vessel sampler is located within bottom section of a conveyor head hopper. The tilting movement of the cutter is actuated by a side-located pneumatic cylinder or electric actuator (Figure 332).

The width of the slot of the cutter is 20 mm to 50 mm depending on the particle maximum diameter.

The relatively large volume of the cutter (10 kg to 20 kg) enables it to contain a complete cut of the material flow. In the rest position, the cutter slot is covered by flexible rubber seals.

The primary sample is directed to the splitter/divider.

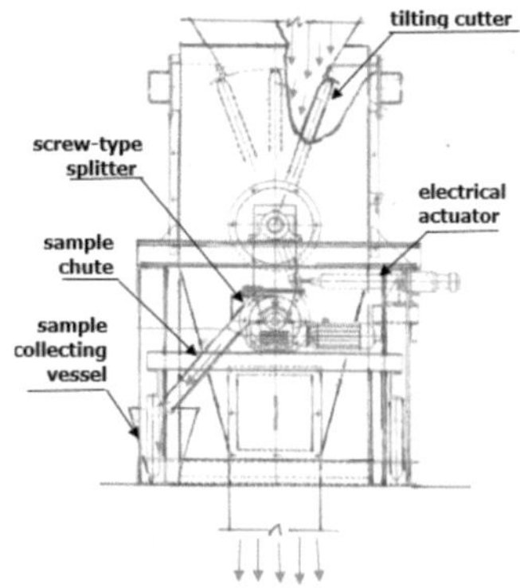

Figure 332. Tilting-type sampler with built-in splitter (see Figure 334).

7.2.2.4 Screw-type Samplers

Screw samplers (Figure 333) are usually used in the pharmaceutical industry for taking samples for chemical composition analysis (not for particle-size distribution analysis) from a low-capacity material flow.

The sampler consists of a slotted tube 50 mm in diameter with an inner screw.

When the sampler moves forward in the flow, the material penetrates through the slots in and is transported by the inner screw into discharge tube.

The screw can be powered by a small gear-motor or manually.

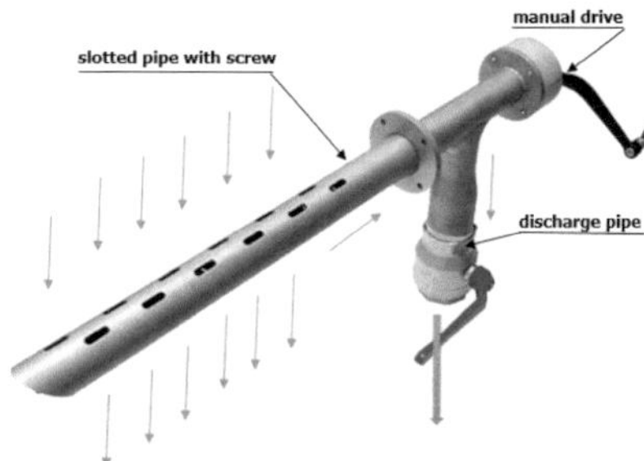

Figure 333. Manually powered screw sampler (Siebtechnik GmbH).

7.3 Special Samplers

7.3.1 Special Samplers with Built-in Splitters (SBS)

The sampler (Figure 332, Figure 334) can be installed under the lower hopper of a belt conveyor and consists of a hopper, a slotted tilting cutter (with an open slot 30 mm wide and up to 20 kg receiving capacity), and a screw-type splitter developed by the author.

The outer pneumatic cylinder (or electrical actuator) forces the pivoted cutter to tilt between two rest positions (+15° and −15°) within the hopper.

During sample-taking movement, the cutter is tilted in one direction only, cuts the material flow, and collects the primary sample. The sample material falls directly into the splitter. The splitter is a short (having only five spirals) screw. A small spoon (containing 10 g to 12 g) is attached to the shaft of the screw by means of a lever and rotates together with the screw.

Each screw rotation, a small amount of the primary sample material is scooped with the spoon and is discharged (in accordance with the calculated centrifugal trajectory) through special curved chute into collecting container.

Thus, during each revolution of the screw, the new portion of the primary sample material (the first 2.5 strokes of the screw have a smaller diameter and filled 100%) is advanced to the spoon and the spoon scoops a small amount (sub-secondary-sample) from each new portion to gather the representative secondary sample in the container.

The outer diameter of the spoon is equal to the screw's spiral outer diameter and it is smaller than the diameter of the housing (delta $\Delta = 5$ mm), so the spoon does not rub the housing and each time takes sub-samples from the new portion of the bulk material "supplied" by the screw.

The reject of the sample material is conveyed by the same screw back into the hopper, so no special equipment is required for disposal of the reject.

During rest periods, the slot of the tilting cutter is covered by rubber seals to prevent penetration of dust.

Comparative tests were conducted between the results of this sampler and the samples taken from stopped conveyor and showed very similar results.

The *advantage* of the SBS sampler is its simplicity, reliability, and representativeness.

This sampler successfully operated in port Eilat.

A similar sampler, manufactured by M&W JAWO Handling Co., Denmark, consists of a slotted vessel, fed through the central inlet and rotating inside the casing around the vertical axis, and a built-in splitter.

To take the primary sample, the container makes one rotation of 180° and cuts the flow of the bulk material by the slotted vessel. The primary representative sample falls through outlet pipe into a fixed chute while several rotating baffles (plates) throw the sub-samples from the falling flow of the primary sample material into collecting container (Figure 333.1).

The disadvantage of the sampler can be considered a method of obtaining secondary samples, when the rotated baffles (plates) impact the flow of particles of the primary sample, throwing out particles (and, naturally, breaking some of them) into bypass pipe as a secondary sample. In contrast with this splitter, the screw splitter, described above (Figure 332, Figure 334), takes sub-samples by cutting the slow-moving layer of the primary sample material without breaking the particles.

Another drawback of M&W splitter is the problem of disposal of the reject bulk material.

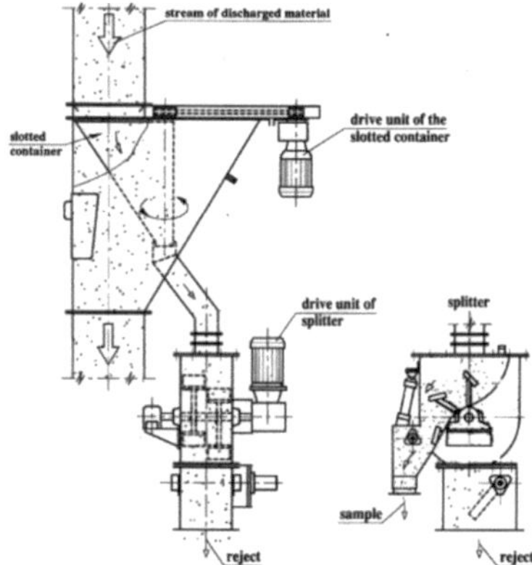

Figure 333.1. Combined slotted sampler with automatic splitter.

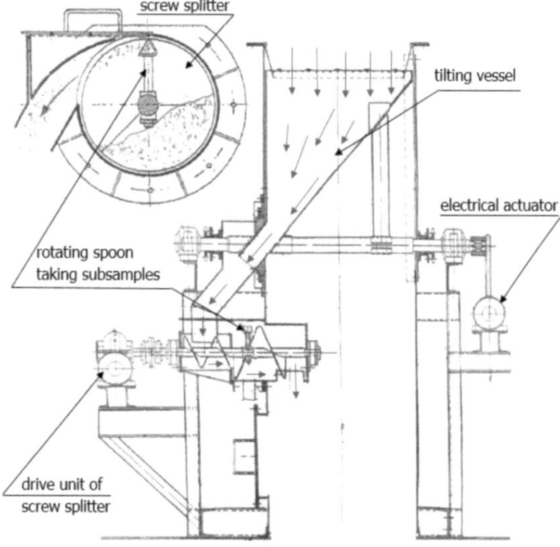

Figure 334. Combined tilting cutter sampler with built-in screw splitter.

*Advantages of the SBS sampler (*Figure 332, Figure 334*):*

1. The full cross-sectional cut of the flow, taken during the one-directional tilting of the open-slotted vessel, provides a representative primary sample without overflowing the vessel.
2. During rest periods, the slotted vessel is out of the material flow and its inlet slot is covered by a rubber seal.
3. The splitter is built in with the sampler, and rejected material is directed back into the hopper by the screw of the splitter. Takeaway conveyors or elevators for disposal of the rejected material are not required.
4. The screw of the splitter rotates all the time to clean the device between sample-taking operations.
5. The number of small secondary subsamples allows the final secondary sample to be representative. For example, author's calculations showed that 7÷8 small subsamples, at 50% representativeness each, enough to provide one 95% representative secondary sample.

Disadvantages of the sampler:

1. The device is relatively large; therefore, its installing should be designed from the very beginning of the project.
2. In a humid environment, the small increments of hydroscopic bulk material should be kept hermetically closed.

7.4 Splitters

As already noted, the primary sample, taken from a high-capacity conveyor, should be relatively large (10 kg to 20 kg) to be the representative one, but for analysis the material in a lab (chemical analysis, particle-size distribution) a relatively small amount of the material (100 g to 200 g) is required. So, the primary sample must be splitted or divided, but in such a way that the small, secondary sample must be the representative also.

There are two main methods of primary sample dividing/splitting:

- manual splitting
- automatic splitting

Manual Splitting

Manual splitting is the process of successive halving the primary sample by manually pouring out the material through a special *X*-shaped device (or riffle divider) that has slots arranged in opposite directions to transfer 50% of the sample into collecting vessel no. 1 and 50% into vessel no. 2 (Figure 335, Figure 336). After several successive halvings, the small representative sample can be sent to the lab.

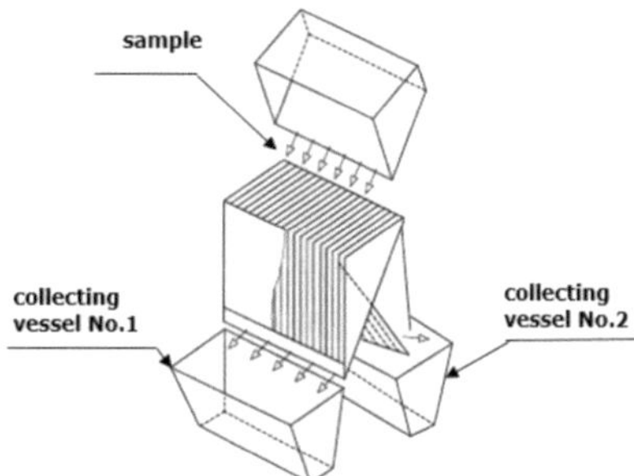

Figure 335. Principle of the manual splitter operation: successive halving of the primary sample many times to get a small representative secondary sample.

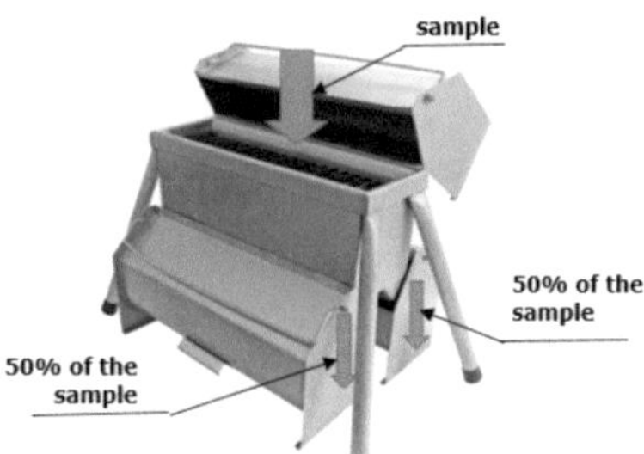

Figure 336. The laboratory manual splitter (Siebtechnik GmbH).

Advantage of manual splitters:

1. Simple, reliable, manually controlled splitting.

Disadvantages of manual splitters:

1. Slow, time-consuming procedure.
2. The device is recommended to clean after each halving.

Automatic Splitting via Rotary Pipe Divider

The operational principle of a rotary pipe divider (Figure 337) is that the material flow is distributed by the sloped rotating pipe within a conical casing. The cone has special opening closed by a slide plate. When the opening opens, and the pipe is in front of the opening, the bulk material (sample) from the rotating pipe enters the sample chute. When the opening closes, the pipe continues rotate releasing the bulk material inside the casing. Dividing ratio of the sample depends on regularity of the opening and closing of the slide plate.

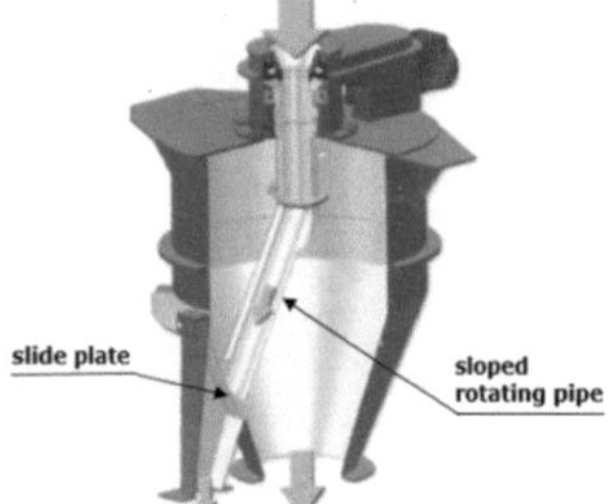

Fig. 337 Rotary pipe divider (Siebtechnik GmbH).

LITERATURE

1. Jenike, A.W., "Storage and Flow of Solids", University of Utah Engineering Experiment Station, Bulletin No 123, Nov., 1964.
2. W. Reisner, M.v.E. Rothe, "Bins and Bunkers for Handling Bulk Materials", 1971.
3. CEMA "Belt Conveyors for Bulk Materials", 6th Edition.
4. "Industrial Ventilation", 26th Edition, 2007.
5. ISO 5048: 1989, Continuous mechanical handling equipment – Belt conveyors with carrying idlers – Calculation of operating power and tensile forces.
6. DIN 2210: 2011-12, Continuous conveying – Bulk conveyors for loose bulk materials – Basis for calculations.
7. J. Fruchtbaum, Bulk Material Handling Handbook, 1988.
8. Belt Conveyor Fatal Nip Hazards, HSE SPC/Tech/ SI/11. National Safety Council, Data Sheet 569, 2016
9. MSHA Guide to Equipment Guarding, 2003.
10. CEMA SBR-001 (2004) "Safety Best Practices Recommendations".
11. "Foundations of Safe Bulk-Materials Handling", Martin Eng., USA
12. DIN 22102-1 "Conveyor belt with textile plies for bulk goods".
13. ISO 6502: 2016 "Rubber".
14. ISO 7623: 2015 "Steel cord conveyor belts". ISO 7623: 2015 "Steel cord conveyor belts".
15. CEMA 576 "Classification for conveyor belt cleaners".
16. ANSI/CEMA 501.1, "Welded steel conveyor pulleys", 2015.
17. ANSI/CEMA 501.1, "Welded steel wing pulleys", 2015.
18. CEMA Standard 502-2004 "Troughing and Return Idlers".
19. Sandvik HR 310, Carry idlers.
20. Rex Conveyor Idlers, Rexnord Catalog.
21. Nordel, L.K., Ciozda, Z.P., Transient belt Stresses during Starting and Stopping: Elastic Response Simulated by Finite Element Methods. Bulk Solids Handling 4/1, 1984, pp. 93-98.

22. CEMA "Bucket Elevators Book", 1st Ed., 2017

23. Bucket Elevators, Aumund, Germany.

24. ANCI/CEMA STANDARD No. 350, 2009, "Screw Conveyors for Bulk Materials".

25. "Screw Conveyor Catalogue and Engineering Manual", Continental

26. Drag Chain Conveyors, Aumund, Germany.

27. KWS Design Standard, "Drag Conveyor Engineering Guide".

28. D. Mills, Pneumatic Conveying Design Guide, 2nd ed., 2004.

29. A.T. Agarwal, Theory and Design of Dilute Phase Pneumatic Systems, Powder, Vol. 17, No. 1, 2005.

30. K.C. Williams, Dense Phase Pneumatic Conveying of Powders, 2008.

31. A.N.M. van der Biji, "Bulk Handling Equipment of Grains", 2013.

32. Handbook of Air Pollution, CRC Inc., 1995.

33. EPA-452/F-03-017, Air Pollution Control Technology.

34. D. Sanders, Venturi Scrubbers, SLY Inc., USA, Processing, 2016.

35. R. Halton, Wet Scrubbers for dry Dust? Absolutely, SLY Inc., USA.

36. Fertilizer International, 465, 2015.

37. J.W. Merks, Sampling and Weighing of Bulk Solids, 1985.

38. "Code of Practice for the Safe Loading and Unloading of Bulk Carriers" (13-98-IMO).

39. ISO 5049-1, 1994, Mobile continuous bulk handling equipment.

40. ISO 11648-1:2008, Statistical aspects of sampling from bulk materials.

41. IACS -International Association of Classification Societies Ltd., 2017.

Printed in the United States
By Bookmasters